"The ultimate test of a state's ability to survive comes from crises, whether a natural disaster or a man-made political crisis. For all those interested in the survival of the incumbent regime in China, this unique and informative book will be required reading."

Bruce J. Dickson, *George Washington University*

"In thirty years of rapid social and economic development, China's crisis management and crisis avoidance have been exemplary. This volume provides an invaluable guide to crisis development and response in the People's Republic."

David S. G. Goodman, *University of Sydney*

"A remarkably timely study, which addresses key issues such as government capacity, leadership abilities, regime legitimacy, system survivability, all leading to the perennial question - will the People's Republic be able to manage all its crises and continue long into the future?"

David Zweig, *Hong Kong University of Science and Technology*

China's Crisis Management

The continuation of China's successful rise depends considerably on the capacity of the Chinese government to prevent and manage a wide range of potential and actual crises, which could, if mishandled, have serious adverse consequences for China. These potential crises are both domestic – where the example of the collapse of the Soviet Union is well understood and remembered in China – and, increasingly transnational, as a result of China's ever closer involvement in the global system. This book presents a comprehensive overview of crisis management in China, and examines China's mode of managing economic, political and military crises, as well as natural disasters, ethnic-minority issues, environmental and public health problems. In each area it considers the nature of potential crises and their possible effects, and the degree to which China is prepared to cope with crises.

Jae Ho Chung is Professor of International Relations and Director of the Institute for China Studies at Seoul National University, Korea. His most recent books include *China's Local Adminstration* (Routledge, 2010).

China Policy Series

Series editor: Zheng Yongnian

China Policy Institute, University of Nottingham, UK

1. **China and the New International Order**
   Edited by Wang Gungwu and Zheng Yongnian
2. **China's Opening Society**
   The non-state sector and governance
   Edited by Zheng Yongnian and Joseph Fewsmith
3. **Zhao Ziyang and China's Political Future**
   Edited by Guoguang Wu and Helen Lansdowne
4. **Hainan – State, Society, and Business in a Chinese Province**
   Kjeld Erik Brodsgaard
5. **Non-Governmental Organizations in China**
   The rise of dependent autonomy
   Yiyi Lu
6. **Power and Sustainability of the Chinese State**
   Edited by Keun Lee, Joon-Han Kim and Wing Thye Woo
7. **China's Information and Communications Technology Revolution**
   Social changes and state responses
   Edited by Xiaoling Zhang and Yongnian Zheng
8. **Socialist China, Capitalist China**
   Social tension and political adaptation under economic globalisation
   Edited by Guoguang Wu and Helen Lansdowne
9. **Environmental Activism in China**
   Lei Xei
10. **China's Rise in the World ICT Industry**
    Industrial strategies and the catch-up development model
    Lutao Ning
11. **China's Local Administration**
    Traditions and changes in the sub-national hierarchy
    Edited by Jae Ho Chung and Tao-chiu Lam
12. **The Chinese Communist Party as Organizational Emperor**
    Culture, reproduction and transformation
    Zheng Yongian
13. **China's Trade Unions – How Autonomous are They?**
    Masaharu Hishida, Kazuko Kojima, Tomoaki Ishii and Jian Qiao
14. **Legitimating the Chinese Communist Party since Tiananmen**
    A critical analysis of the stability discourse
    Peter Sandby-Thomas
15. **China and International Relations**
    The Chinese view and the contribution of Wang Gungwu
    Zheng Yongnian
16. **The Challenge of Labour in China**
    Strikes and the changing labour regime in global factories
    Chris King-chi Chan
17. **The Impact of China's 1989. Tiananmen Massacre**
    Edited by Jean-Philippe Béja
18. **The Institutional Dynamics of China's Great Transformation**
    Edited by Xiaoming Huang
19. **Higher Education in Contemporary China**
    Beyond expansion
    Edited by W. John Morgan and Bin Wu
20. **China's Crisis Management**
    Edited by Jae Ho Chung

China's Crisis Management

Jae Ho Chung

Routledge
Taylor & Francis Group
LONDON AND NEW YORK

This edition published 2012
by Routledge
2 Park Square, Milton Park, Abingdon, Oxon, OX14 4RN

Simultaneously published in the USA and Canada
by Routledge
711 Third Avenue, New York, NY 10017

Routledge is an imprint of the Taylor & Francis Group, an informa business

British Library Cataloguing in Publication Data
A catalogue record for this book is available from the British Library

Library of Congress Cataloging-in-Publication Data
China's crisis management / edited by Jae Ho Chung.
p. cm.—(China policy series; 20)
Includes bibliographical references and index.
1. Crisis management—China. 2. Emergency management—China. 3. Disaster relief—China. 4. China—Politics and government—2002– I. Chong, Chae-ho, 1960– II. Title. III. Series: China policy series; 20.
HV551.5.C6C56 2011
363.34′80951—dc22
2011011395

ISBN 978–0–415–67780–6 (hbk)
ISBN 978–0–203–80357–8 (ebk)

Typeset in Times New Roman
by Book Now Ltd, London

MIX
Paper from responsible sources
FSC® C004839

Printed and bound in Great Britain by the MPG Books Group

Contents

Illustrations

Figures

Tables

Contributors

Gang Chen is Research Fellow at the East Asian Institute, National University of Singapore. He is the author of *The Kyoto Protocol and International Cooperation against Climate Change* (Xinhua Press, 2008) and *Politics of China's Environmental Protection: Problems and Progress* (World Scientific, 2009).

Jae Ho Chung is a professor of international relations and Director of the Institute for China Studies at Seoul National University in Korea. He is the author or editor of eleven books, including *Central Control and Local Discretion in China* (Oxford University Press, 2000), *Between Ally and Partner: Korea–China Relations and the United States* (Columbia University Press, 2007), and *China's Local Administration* (Routledge, 2010).

Hongyi Lai is Associate Professor at the School of Contemporary Chinese Studies, University of Nottingham, UK. His publications include seven books and articles in such journals as *China Quarterly, China Journal, Modern China*, and *Third World Quarterly*. His most recent book is *The Domestic Sources of China's Foreign Policy* (Routledge, 2010).

Colin Mackerras is Professor Emeritus at Griffith University, Australia. He has written widely on China and his books include *China's Ethnic Minorities and Globalisation* (2003), and the four-volumed *Ethnic Minorities in Modern China, Critical Concepts in Asian Studies*, (2011).

Richard P. Suttmeier is Professor of Political Science, Emeritus, at the University of Oregon. He has written widely on scientific and technological development in China. His recent work can be found at http://china-us.uoregon.edu/papers.php.

Tuosheng Zhang is Chairman of the Academic Committee and Director of the Center for Foreign Policy Studies at the China Foundation for International and Strategic Studies, China.

Wei Zhang is a visiting fellow of the Brookings Institution. He was the Director of Cambridge Research Unit for Chinese Economy at East Asian Institute and Lecturer in Chinese Economy at the Faculty of Asian and Middle Eastern Studies, University of Cambridge from 2000 to 2010.

Preface

Every state has the ultimate imperative: survival. Threats to their survival may come either from outside or from within. In an era of globalization where borders and nationalities are much less relevant than in the past, the domestic–external distinction is getting increasingly blurred. That is to say, the issue of national security is as much domestic as it is external, and epidemics/pandemics are as much foreign as they are indigenous.

China, in the midst of arduous reforms of transition and all-directional opening, is facing a wide array of new and complex threats. Its leadership has been highly concerned with the eruption of systemic crises and, therefore, has emphasized the crucial importance of stability. Considering the fact that Chinese history is replete with dynastic collapses and territorial disintegration, such concern is by no means ill-founded. The Tiananmen demonstrations, the *Falungong* incident, the EP-3 clash, the SARS crisis, falling real-estate prices, and the tumultuous revolts in Tibet and Xinjiang areas are only the notable examples of the problems that the Chinese government is facing every now and then on the mainland.

Despite the rising frequency and scale of these wide-ranging problems, China has thus far successfully brushed off the prospect of a system-wide crisis. What enables Beijing to maintain its power and efficacy to rule the continent with 1.3 billion people? In order to answer this crucial question, this volume presents comparative sectoral analyses of China's mode of managing (semi-)crises. More specifically, each of the seven chapters examines the history, organizations, learning effects, and norms of managing crises in economic, political, military, ethnic-minority, pandemic/epidemic, environmental, and natural disaster-related domains. I am grateful to all the contributors for their time, effort, and insight.

This book is based upon the papers presented at the international conference entitled "China's Crisis Management: Evolving Norms and Learning," held in Seoul during May 26–27, 2010. I, as the organizer and editor, would like to acknowledge the support and encouragement from the Institute for China Studies at Seoul National University. Thanks are also due to the National Research Council for Economics, Humanities and Social Sciences and its president, Dr Caeone Kim, for co-organizing the conference. Hon S. Chan, Young Nam Cho, and Keun Lee are also acknowledged for their role as meticulous discussants.

I owe a special thank-you to Yongnian Zheng, the series editor, and Peter Sowden, editor at Routledge, for their support and encouragement. Thanks also go to Myungjong Jun, chief assistant at the Institute for China Studies, for his effective administration of the conference as well as of the editing task for this book. Finally, I would like to express my sincere gratitude to Hye Kyung and Jean for their love and prayer.

Jae Ho Chung
January 2011

1 China's anti-crisis macro-economic management

Effectiveness and limitation

Wei Zhang

Economic crisis generally refers to a phenomenon of a sharp downturn of national output. It may take some specific forms, or more likely a combination of various forms, such as stagnation, recession, or depression in terms of gross domestic product (GDP) growth, hyperinflation or deflation in terms of price level, high unemployment rate, huge government debts, persistent trade deficit, etc.

Anti-crisis management has become a significant part of macro-economic management of almost all governments in the modern world. No politician, even those who strongly advocate for libertarian economic philosophy, will adopt absolute laissez-faire policies in dealing with economic development, especially when a country faces an economic crisis. The rationale behind this phenomenon is pretty simple; economic crises can easily drive the existing political leaders out of office in democratic countries and destabilize political and social orders, or even cause regime changes in non-democratic countries.[1] The great efforts that most governments in the world have made to fight the global financial crisis since 2008 clearly demonstrate their consensus on the importance of the state's role in macro-economic management.

This chapter attempts to explore the features, mechanisms, effectiveness, and constraint of China's anti-crisis macro-economic management system. This research is mainly based on China's macro-economic management, which aims to offset the negative shocks of the current global financial crisis; at the same time, it also makes reference to China's experience of handling the Asian Financial Crisis in 1998 and other economic problems.

As is well known, the global economy went through a period of unprecedented financial instability in 2008–09, accompanied by the deepest economic recession and collapse in international trade in many decades. This crisis, which originated in advanced-country markets in 2007, has spread rapidly across the whole world, engulfing almost every part of the global economy including advanced economies, emerging markets, and low-income countries alike. Many regard it as "the worst economic slowdown since the Great Depression."[2] Two years later, the whole world is still on a slow and uncertain process of recovery.

After thirty years' active engagement in the economic globalization, China could not and did not escape the reach of the latest global economic storm; however, the Chinese economy has shown remarkable resilience in the face of

Table 1.1 Global GDP growth, 2008–10 (% change year-on-year)

	2008	*2009*	*2010*[a]
World output	3	−0.8	3.9
Advanced economies	0.5	−3.2	2.1
United States	0.4	−2.5	2.7
Euro area	0.6	−3.9	1
Germany	1.2	−4.8	1.5
France	0.3	−2.3	1.4
Italy	−1	−4.8	1
Spain	0.9	−3.6	−0.6
Japan	−1.2	−5.3	1.7
United Kingdom	0.5	−4.8	1.3
Canada	0.4	−2.6	2.6
Other advanced economies	1.7	−1.3	3.3
Newly industrialized Asian economies	1.7	−1.2	4.8
Emerging market and developing economies	6.1	2.1	6
Africa	5.2	1.9	4.3
Saharan Africa	5.6	1.6	4.3
Central and eastern Europe	3.1	−4.3	2
Commonwealth of Independent States	5.5	−7.5	3.8
Russia	5.6	−9	3.6
Excluding Russia	5.3	−3.9	4.3
Developing Asia	7.9	6.5	8.4
China	9.6	8.7	10
India	7.3	5.6	7.7
AS5	4.7	1.3	4.7
Middle East	5.3	2.2	4.5
Western Hemisphere	4.2	−2.3	3.7
Brazil	5.1	−0.4	4.7
Mexico	1.3	−6.8	4

Source: International Monetary Fund, *World Economic Outlook Update*, January 2010, p. 2.

Note
a 2010 data are projections.

the crisis. Compared to the world economy as a whole and other large economies in particular, China's economic performance has been impressive. While the world GDP dropped by 0.8 percent, with advanced economies dropping 3.2 percent, China managed to achieve economic growth of 8.7 percent in 2009 (see Table 1.1).[3] As to China's foreign trade, decline in the demand for China's products in the international market caused a big fall in China's trade. The total value of China's exports and imports in 2009 decreased by 16 and 11 percent, respectively, but the decrease margin of China's external trade was much smaller than other main trade powers in the world (see Tables 1.2 and 1.3).[4]

Moreover, China has taken the crisis as an opportunity to increase its economic weight and political influence in the world. During the period of the world economic recession, China's share in the global economy increased from 7.7 to 8.5 percent from 2008 to 2009; with regard to international trade, China has replaced

Table 1.2 Leading merchandise exporters, 2009 (US$bn and %)

	Value	*Share*	*Change*	*Rank*
China	1,202	9.6	−16	1
Germany	1,121	9	−22	2
United States	1,057	8.5	−18	3
Japan	581	4.7	−26	4
Netherlands	499	4	−22	5
France	475	3.8	−21	6
Italy	405	3.2	−25	7
Belgium	370	3	−22	8
Korea, Republic of	364	2.9	−14	9
United Kingdom	351	2.8	−24	10

Source: World Trade Organization, *World Trade 2009, Prospects for 2010*, p. 11.

Table 1.3 Leading merchandise importers, 2009 (US$bn and %)

	Value	*Share*	*Change*	*Rank*
United States	1,604	12.7	−26	1
China	1,006	8	−11	2
Germany	931	7.4	−21	3
France	551	4.4	−22	4
Japan	551	4.4	−28	5
United Kingdom	480	3.8	−24	6
Netherlands	446	3.5	−23	7
Italy	410	3.2	−26	8
Hong Kong	353	2.8	−10	9
Belgium	351	2.8	−25	10

Source: World Trade Organization, *World Trade 2009, Prospects for 2010*, p. 11.

Germany as the largest exporter and the second largest importer in the world. The shares of China's exports and imports in 2009 accounted for 9.6 and 8 percent of the world total.[5] China's foreign reserve continued to increase from 1528.2 billion US dollars at the end of 2007 to 2,399.3 billion US dollars at the end of 2009, up by 57 percent during the crisis period (see Table 1.4 and Figure 1.1). During the period from July 2008 to March 2010, China's official holding of US treasury securities increased from 518.7 billion to 895.2 billion US dollars, and now it is the No. 1 US treasury holder in the world.[6] In acknowledgement of China's growing economic and political influence, the World Bank and International Monetary Fund have raised China's voting rights.[7]

This is not the first time that China has survived the shocks of the international economic crises. Since it started market-oriented economic reform and embarked on economic globalization in 1978, China has successfully withstood dramatic negative impacts of two serious economic crises. In addition to the current global financial crisis, China survived the Asian Financial Crisis in

Table 1.4 Changes in China's foreign reserve during the global crisis

Month	*Billion US$*	*Month*	*Billion US$*
2007–12	1,528.20	2009–2	1,912.10
2008–1	1,589.80	2009–3	1,952.70
2008–2	1,647.10	2009–4	2,008.90
2008–3	1,682.20	2009–5	2,089.50
2008–4	1,756.70	2009–6	2,131.60
2008–5	1,797.00	2009–7	2,174.60
2008–6	1,808.80	2009–8	2,210.80
2008–7	1,845.20	2009–9	2,272.60
2008–8	1,884.20	2009–10	2,328.30
2008–9	1,905.60	2009–11	2,388.80
2008–10	1,879.70	2009–12	2,399.20
2008–11	1,884.70	2010–1	2,415.20
2008–12	1,946.00	2010–2	2,424.60
2009–1	1,913.50	2010–3	2,447.10

Source: State Administration of Foreign Exchange, People's Republic of China, at http://www.chinability.com/Reserves.htm.

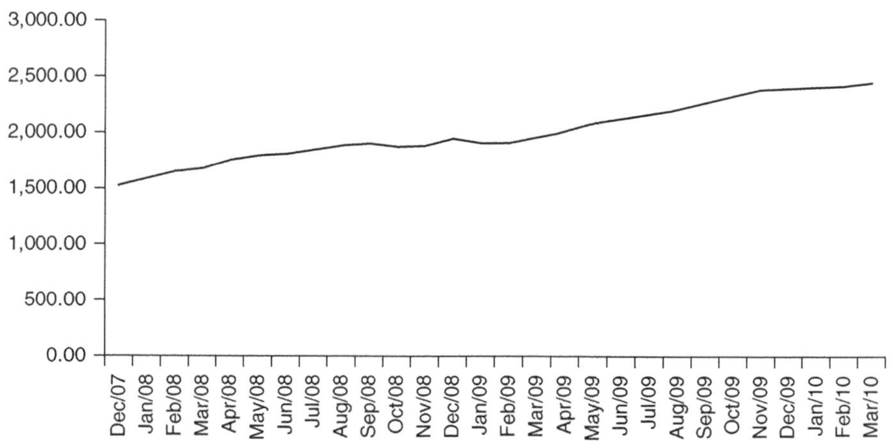

Figure 1.1 Changes in China's foreign reserve during the global crisis (bn US$).

1997–98. During that period, all newly industrialized Asian economies (both the four dragons and four tigers) and Japan suffered sudden and large financial problems, trade collapse, and economic decline. Although China's GDP experienced a short period of trade decline in 1998 and a slowdown of GDP growth from 9 percent in 1997 to 8 percent in 1998 (see Table 1.5 and Figure 1.2), it recovered from the impacts of the Asian Financial Crisis relatively quickly. And by any standards, the growth rate of 8 percent should be regarded as a very good economic performance, let alone that it happened during the period of a serious regional economic crisis.

Table 1.5 GDP Growth during the Asian Financial Crisis (%)

	1996	*1997*	*1998*	*1999*
China	10	9	8	7
Hong Kong	4	5	−5	3
Indonesia	8	5	−13	1
Korea, Rep.	7	5	−7	9
Malaysia	10	7	−7	6
Philippines	6	5	−1	3
Singapore	8	9	−1	6
Thailand	6	−1	−11	4
Japan	3	2	−1	0

Source: World Bank, *World Development Indicators 2005.*

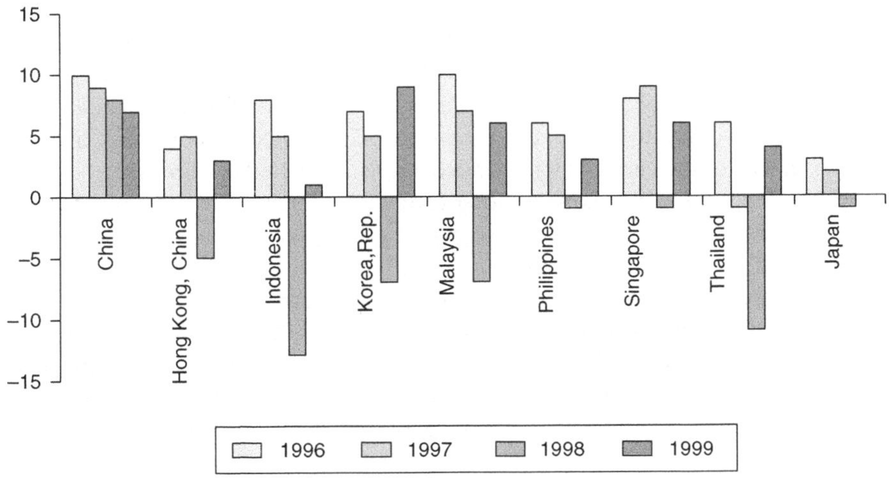

Figure 1.2 GDP growth during the Asian Financial Crisis.

Source: World Bank, *World Development Indicators 2005* (Washington, DC: World Bank Publications 2005).

An important explanation of China's outstanding economic performance during the regional and global economic crises is the government's macro-economic policies of offsetting negative impacts of external shocks. This very fact invites some interesting intellectual questions, such as, what kind of policies did the Chinese government adopt in dealing with the economic crises? What are the main mechanisms of China's anti-crisis policy implementation? If those policies work well in offsetting external shocks in the short term, do they have any negative impacts on the Chinese domestic economy in the long term? To answer those questions is a main incentive of this chapter.

The chapter is arranged as follows: after this introduction, I will take a brief look at the impacts of the current global financial crisis on the Chinese economy; analyze China's policy response to the external shocks and effectiveness of its anti-crisis policies; describe some key necessary conditions for the effective implementation of those anti-crisis polices; and finally, explore the post-crisis macro-economic management faced by China.

Impacts of the global financial crisis on the Chinese economy

The ongoing global financial troubles have long and deep roots; they started to spread to the real economy in the summer of 2008, and hit the Chinese economy in the last quarter of 2008. After three decades of economic reform, the Chinese economy became highly dependent on international trade and foreign investment.[8] Because of extensive exposure to the regional and global economies, any large shocks from other parts of the world economy inevitably have tremendous negative impacts on China's economic growth. China's export growth slowed down rapidly during the Asian Financial Crisis from 21 percent in 1997 to 0.5 percent in 1998;[9] and during the current global crisis from 26 percent in 2007 to − 16 percent in 2009.[10] According to an empirical analysis by Ligang Liu, a decline in economic growth of one percentage point in the USA, the European Union, and Japan is likely to lead to a decline in growth of 0.73 percentage points in China.[11]

The shocks of the current global financial crisis affected the Chinese economy mainly through three channels: the first channel was a sudden and large drop in the international demands for Chinese export goods; the second channel was the shrinkage of foreign direct investment (FDI) in China; the third channel was the contagious decline in confidence among China's private investors.

The decline in world demand for China's exports is the most direct and important channel through which the global financial crisis hit the Chinese economy. The growth rate of China's exports in October 2008 compared to the same period in the previous year was at a very high level of 19.2 percent, but it suddenly fell to − 2.2 percent in November. The decline in China's exports continued even faster in the following months until November 2009. It demonstrates a sharp drop if we make a comparison between the export performance in the first three quarters of 2008 and that in the first three quarters of 2009. The average growth rate of exports in the first nine months of 2008 was 22.2 percent, and the average growth rate of exports in the first nine months of 2009 was − 21.4 percent (see Table 1.6).

The global financial crisis also hit FDI in China heavily. Since the 1990s, the FDI had sustained high growth rates. This phenomenon continued up to October 2008 when the increase in FDI was 35.6 percent on a year-to-year basis; however, there appeared a dramatic decrease of − 36.52 percent in November 2008, and the decline in FDI lasted for nine months until September 2009. The average monthly growth rate of the first nine months in 2008 is 42.9 percent, but that rate decreased rapidly to − 12.8 percent in the first nine months of 2009 (see Table 1.7).

The reduction in foreign investment is partly because of the credit crunch, which made it difficult for potential foreign investors to get loans from foreign banks. It is also because of the rapid shrinkage of international demand for China's export goods, because the majority of foreign-invested firms are producers of

Table 1.6 Changes in China's exports (%)

Month	*Change*	*Month*	*Change*	*Month*	*Change*
2008–1	26.50	2009–1	−17.50	2010–1	21.00
2008–2	6.30	2009–2	−25.70	2010–2	45.00
2008–3	30.30	2009–3	−17.10	2010–3	24.00
2008–4	21.80	2009–4	−22.60	2010–4	30.50
2008–5	28.10	2009–5	−26.40		
2008–6	17.20	2009–6	−21.40		
2008–7	26.80	2009–7	−23.00		
2008–8	21.00	2009–8	−23.40		
2008–9	21.50	2009–9	−15.20		
2008–10	19.20	2009–10	−13.90		
2008–11	−2.20	2009–11	−1.20		
2008–12	−2.80	2009–12	17.70		

Source: Collected by the author based on the *Monthly Reports of China's Foreign Trade*, published by the Ministry of Commerce.

Table 1.7 Changes in China's realized FDI (%)

Month	*Change*	*Month*	*Change*	*Month*	*Change*
2008–1	109.78	2009–1	−32.67	2010–1	7.79
2008–2	38.31	2009–2	−15.81	2010–2	4.86
2008–3	39.60	2009–3	−9.50	2010–3	12.08
2008–4	42.34	2009–4	−22.51	2010–4	11.28
2008–5	37.95	2009–5	−17.81		
2008–6	33.28	2009–6	−6.76		
2008–7	36.30	2009–7	−35.71		
2008–8	20.38	2009–8	7.00		
2008–9	27.85	2009–9	18.93		
2008–10	35.06	2009–10	5.70		
2008–11	−36.52	2009–11	31.90		
2008–12	−5.730	2009–12	103.00		

Source: Collected by the author based on *Monthly Reports of China's Foreign Direct Investment*, published by the Ministry of Commerce.

export products. The export goods produced by foreign-invested firms account for more than 55 percent of China's total exports (see Table 1.8).[12]

Following the first two channels of impacts of the world financial crisis, the loss of confidence of domestic entrepreneurs has become a serious problem. According to an official survey on the confidence of investors and senior management of enterprise, which was conducted by China's National Bureau of Statistics, in the fourth quarter of 2008 the index of the confidence dropped by about 30 per cent compared to the third quarter of the same year.

The decline in exports, together with weakening confidence, inevitably and very quickly affected Chinese manufacturing industries. The average monthly

Table 1.8 Share of foreign-invested firms in total exports (US$bn and %)

Year	*Total exports*	*Exports by FIFs*	*Share of FIFs*
2005	762.0	444.2	58.3
2006	969.0	563.8	58.2
2007	1218.0	695.5	57.1
2008	1428.5	790.6	55.3
2009	1201.6	672.2	55.9

Source: Collected by the author from *Data of Major Statistics Monthly*, published by China Customs Statistics Information.

Table 1.9 Monthly changes in industrial value added (%)

Month	*Change*	*Month*	*Change*
2008–1	18.9	2009–1	4.8
2008–2	15.4	2009–2	11
2008–3	17.8	2009–3	8.3
2008–4	15.7	2009–4	7.3
2008–5	16	2009–5	8.9
2008–6	16	2009–6	10.7
2008–7	14.7	2009–7	10.8
2008–8	12.8	2009–8	12.3
2008–9	11.4	2009–9	13.9
2008–10	8.2	2009–10	16.1
2008–11	5.4	2009–11	19.2
2008–12	5.7	2009–12	18.5

Source: *Monthly Report of Main Economic Indicators*, published by National Bureau of Statistics.

growth rate of the value added in China's industries on a year-to-year basis decreased from 13 percent in the third quarter to 6.3 percent in the fourth quarter in 2008; it further declined to 4.8 percent in January 2009, reaching the lowest point since the period between 1989 and 1991 (see Table 1.9). The statistical analysis shows that there is significant positive correlation between the changes in China's exports and industrial value added (see Figure 1.3).[13]

A sharp drop in the growth rate of export-related industrial value added had immediately slowed down the growth rate of the GDP. According to research by Yongding Yu, the fall of exports may have cut three percentage points from the GDP growth.[14] A dramatic slowdown in China's economic growth started in the third quarter of 2008; the growth rate decreased from 10.4 percent in the second quarter of 2008 to 6.1 percent in the first quarter of 2009, reaching the lowest level since 1990 (see Table 1.10). The impacts of the slowdown of GDP growth on employment were also grave.[15] The first quarter of 2009 witnessed 1.96 million job losses in China's formal urban employment (not including migrant workers), accounting for 1.6 percent of total formal employment in the urban areas

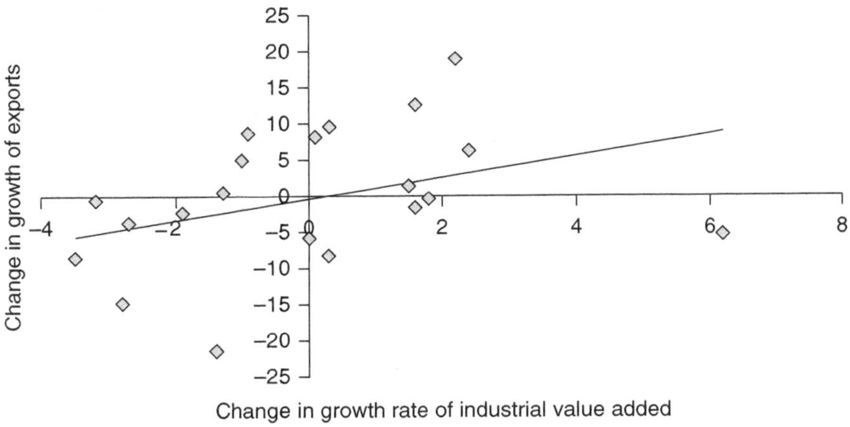

Figure 1.3 Correlation between the first differences of changes in exports and industrial value added (%).

Table 1.10 Quarterly changes in GDP (%)

Quarter	*Change*	*Quarter*	*Change*	*Quarter*	*Change*
2007–1	13.0	2008–1	10.6	2009–1	6.1
2007–2	13.4	2008–2	10.4	2009–2	7.1
2007–3	13.4	2008–3	9.9	2009–3	7.8
2007–4	13.0	2008–4	9.0	2009–4	10.7

Source: *Quarterly Report of Main Economic Indicators*, published by National Bureau of Statistics.

(see Table 1.11). In addition to this, about 20 million migrant workers lost their jobs in urban areas and had to go back to their home villages due to the impacts of the global financial crisis according to Minister Chen Xiwen.[16]

China's policy responses and effectiveness of the policies

Facing the fast slowdown of the national economy, the Chinese government reacted swiftly and decisively. The response from the government to the impacts of the global financial crisis mainly included three parts: while proactive fiscal policies and expansionary monetary policies are among the conventional macro-economic management policy measures, China also quickly strengthened state administrative roles in macro-economic management. The main goals of all policies taken by the Chinese government during the crisis are to minimize the damage caused by the sudden external shocks, and to maintain economic growth and employment at an acceptable level.

Table 1.11 Changes in urban employment (1,000 persons)

Quarter	*Total*	*SOEs*	*COEs*	*Others*
2008–1	−1484	−551	−220	−711
2008–2	1151	343	38	769
2008–3	867	499	−9	378
2008–4	1147	−56	−375	1577
2009–1	−1960	−357	−229	−1373
2009–2	791	204	82	504
2009–3	1233	408	−1	826

Source: *Quarterly Report of Main Economic Indicators*, published by National Bureau of Statistics.

Proactive fiscal policy

In November 2008, the central government introduced a two-year, four-trillion-yuan stimulus package. The total amount involved in the plan is equivalent to about 14 percent of China's GDP in 2008. The central government was responsible for raising 1.18 trillion yuan, 29.5 percent of total stimulus funds. The central money mainly came from newly issued central government bonds and various special funds under the central budget; the rest of the stimulus funds were raised by the local governments, commercial banks, corporate bonds, private funds, etc.[17]

In March 2009, the People's Congress of China approved the central government's new budget for 2009. According to this budget, total government expenditure (central plus local) would reach 7.635 trillion yuan in 2009, up 22.1 percent on the previous year. The total planned government deficit in 2009 was 950 billion yuan (US$139 billion), accounts for 3 percent of the GDP of the same year. The total amount of the deficit is about 8.6 times the deficit in 2008, and the highest budget deficit in six decades. Of this, the planned central government's deficit was 750 billion yuan, 570 billion yuan more than last year. The State Council also for the first time allowed the provincial governments to issue 200 billion yuan worth of government bonds through the Ministry of Finance.

Expansionary monetary policy

Accompanying the proactive fiscal policy, the People's Bank of China (PBOC) has adopted an expansionary monetary policy since 2009. In the first half of 2009, bank credit increased by 7.3 trillion yuan, which exceeded the planned credit target for the whole year. In contrast, the annual increases in bank credit in 2006 and 2007 were 3.18 trillion yuan and 3.63 trillion yuan respectively. The expansionary monetary policy was also reflected in the rapid growth of M1 and M2.[18] From the fourth quarter of 2008, both M1 and M2 started to surge in order to facilitate the government economic stimulates package, and increased 33.2 and 28.4 percent, respectively, in 2009 (see Table 1.12).[19] As a result, the ratio of M2 to GDP reached a historical high.

At the same time, the central bank stopped its sterilization operation with its purchase of US dollars in the exchange market. Previously, after releasing a huge

Table 1.12 Monthly changes in money supply, January 2008–December 2009 (%)

Month	*M2*	*M1*	*M0*	*Year*	*M2*	*M1*	*M0*
2008–1	18.9	20.5	31.2	2009–1	18.7	6.7	12.0
2008–2	17.4	18.9	6.0	2009–2	20.3	10.6	8.3
2008–3	16.2	18.0	11.1	2009–3	25.4	17.0	10.9
2008–4	16.9	18.8	10.7	2009–4	25.9	17.5	11.3
2008–5	18.0	17.7	12.9	2009–5	25.7	18.7	11.2
2008–6	17.3	14.0	12.3	2009–6	28.4	24.8	11.5
2008–7	16.3	13.8	12.3	2009–7	28.4	26.4	11.6
2008–8	15.9	11.3	10.9	2009–8	28.5	27.7	11.5
2008–9	15.2	9.2	9.3	2009–9	29.3	29.5	16.0
2008–10	14.9	8.7	10.6	2009–10	29.5	32.0	14.1
2008–11	14.7	6.6	9.0	2009–11	29.6	34.6	15.0
2008–12	17.8	9.0	12.8	2009–12	28.4	33.2	11.8

Source: Calculated by the author based on "Statistics of Money Supply" from 2007 to 2009, in *Survey and Statistics*, on the website of PBOC (People's Bank of China).

amount of RMB yuan to stabilize the exchange rate against the US dollar, the central bank sold a large amount of central bank bills to mop up the excess liquidity. Since the central bank stopped taking back excess liquidity, the inter-bank money market has been inundated with liquidity. Since the fourth quarter of 2008, the PBOC has almost stopped selling more bills. As a result, liquidity has inundated the inter-bank money market, and once even made the interest rates in the inter-bank market lower than interest on deposits with commercial banks with the same terms of maturity. This excess of liquidity encouraged local governments and state-owned enterprises, which usually have good access to state-owned banks, to expand their borrowing, thus promoting government-led large construction projects.

Strengthening government administrative roles in macro-economic control

In addition to proactive fiscal policy and expansionary monetary policy, the Chinese government also strengthened state administrative power over economic activities, in particular over the expenditure of the four-trillion-yuan stimulus plan. It made a detailed investment plan for the stimulus package, allocating the four trillion yuan funds into seven categories (see Table 1.13):[20]

1. Public infrastructure development: 1.5 trillion yuan, taking up the biggest share, nearly 38 percent of the total package. The projects lined up include railway, road, irrigation, and airport construction.
2. Reconstruction works in regions hit by the magnitude-8 Wenchuan earthquake in May 2008: one trillion yuan, the second largest allocation, about 25 percent of the total fund.

Table 1.13 China's stimulus package

Projects	*Fund (bn yuan)*	*Share (%)*
Public infrastructure	1,500	38
Pearthquake reconstruction	1,000	25
Social welfare	400	10
Technology advancement	370	9
Rural development	370	9
Sustainable development	210	5
Educational & cultural projects	150	4
Total	4,000	100

Source: NDRC (National Development and Reform Commission), 2009.

3 Social welfare plans: 400 billion yuan, about 10 percent of the package, which includes the construction of low-cost housing, rehabilitation of slums, and other social safety-net projects.
4 Rural development: 370 billion yuan, or 9 percent of the total fund. Rural projects in the state plan included building public amenities, resettling nomads, supporting agriculture works, and providing safe drinking water.
5 Technology advancement: 370 billion yuan, 9 percent of the package. It mainly aims at upgrading the Chinese industrial sector, gearing toward high-end production to move away from the current export-oriented and labor-intensive mode of growth. This was in line with the government's latest Blueprint for revitalizing 10 selected industries.
6 To ensure sustainable development, the Chinese government also allocated some 210 billion yuan, or 5.3 percent of the stimulus package for promoting energy saving and poison gas emission cuts, and environmental engineering projects.
7 Last but not least, 150 billion yuan, or 4 percent of the fund, was allocated for educational, cultural and family planning purposes.

In order to maintain economic and social stability, state administrative power over economic activities noticeably increased, it took different forms, including price controls over some important commodities, stricter regulations on state-owned enterprises' or even private firms' decisions over laying off surplus workers, implicit subsidies to export firms through cheaper bank loans and export tax rebate, state interventions on the exchange rate of the RMB yuan against the US dollar, etc.

Being worried about losing competiveness of its export products in the international market where the purchasing power of consumers had already declined due to the economic uncertainty and surge of unemployment rate, the Chinese government gave up the existing exchange policies that began in July 2005 and allowed gradual movement of RMB appreciation based on demand and supply in the market. The official data shows that RMB had appreciated from about 8.27 yuan/1 US$ to around 6.80 yuan/1 US$, or up 21 percent, in the period from July 2005 to July 2008. Since then it has stayed at that level (see Figure 1.4).[21]

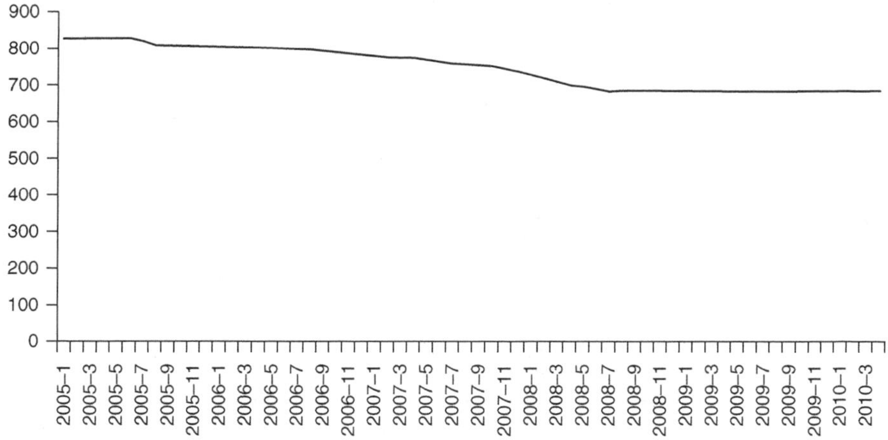

Figure 1.4 Exchange rate of RMB (RMB yuan /100 US$).

Source: Statistics and Data 2010.

Table 1.14 Increases in export tax rebate, August 2008–June 2009

Aug 1, 2008	Textiles and other products
Nov 1, 2008	Textiles and other products
Dec 1, 2008	Labor-intensive products
Jan 1, 2009	Machinery and electronic products
Feb 1, 2009	Textiles and other products
Apr 1, 2009	Light industrial and textile products
Jun 1, 2009	Broad range of products

Source: Ministry of Finance, various official notices of raising tax rebate for part of export commodities.

Another important measure introduced by the central government to stop the downturn trend of exports and promote economic growth was to raise export tax rebate rates. During the period from August 2008 to June 2009, China raised export tax rebate rates seven times, involving more than 8000 export commodities, which accounts for 60 percent of total Chinese export commodities (see Table 1.14).[22] This policy reversed a trend of cutting down export tax rebate rates since 2005, which aimed at adjusting the economic structure and reducing the external dependency of the Chinese economy. The planned fiscal budget for export tax rebate in 2009 was 670 billion yuan, but actual fiscal expenditure on tax rebate reached 800 billion yuan.[23] The goal of raising the export tax rebate rates was to help the export industries to keep their shares in the international market or to slow down the reduction of their shares. Traditionally, there are three policy options that can be used to reach this goal: home currency devaluation or depreciation, increase in government subsidies, and increase in export tax

Table 1.15 Changes in urban fixed-assets investment (%)

Month	*Change*	*Month*	*Change*	*Month*	*Change*
2008–1	24.1	2009–1	26.4	2010–1	27.8
2008–2	24.3	2009–2	26.5	2010–2	26.6
2008–3	25.9	2009–3	28.6	2010–3	26.4
2008–4	25.7	2009–4	30.5		
2008–5	25.6	2009–5	32.9		
2008–6	26.1	2009–6	33.6		
2008–7	27.3	2009–7	32.9		
2008–8	27.4	2009–8	33.0		
2008–9	27.6	2009–9	33.3		
2008–10	27.2	2009–10	33.1		
2008–11	26.8	2009–11	32.1		
2008–12	26.1	2009–12	30.5		

Source: National Bureau of Statistics, urban fixed-assets investment Jan 2008–Mar 2010.

rebate. Since any explicit government subsidies would violate the World Trade Organization's rules, and the depreciation of RMB would also cause serious conflicts between China and its main trade partners, the export tax rebate became the only choice for the Chinese government to rescue its export industry.

Effectiveness of anti-crisis macro-management

China's stimulus plan has brought about positive results for the Chinese economy. Especially, the surge of government-led investment strongly boosted domestic demands and offset the negative impacts caused by the weakened external demands for China's export goods, and thus swiftly prevented a further drop in industrial production, and safeguarded employment for hundreds of millions of urban migrant workers.

A rapid surge of fixed-assets investment has been a main driving force for speeding up economic growth during the period of the crisis. As shown in the National Development and Reform Commission's plans, a large share of the stimulus package went to fixed-assets investment, particularly to those government-led construction projects of the infrastructure of roads, railways, highways, and airports, as well as urban housing and reconstruction of earthquake-hit regions; moreover, a sizable part of the funds nominally allocated for rural development, social welfare, educational and cultural development, technology advancement, and other areas actually also went to fixed-assets investment through various channels.

Table 1.15 shows the growth rates of the fixed-assets investment; the investment started to accelerate quickly by the end of the first quarter of 2009. The growth rate of fixed-assets investment speeded up from 26.2 percent in February to 28.6 percent in March 2009, and accelerated the pace at 2 percentage points a month to 33.6 percent in June 2009, then remained at that level until the end of 2009 (see Table 1.15).

The rapid increase in fixed-assets investment quickly boosted demands for investment-related materials and promoted the fast recovery of the basic industries, including iron and steel, cement, chemicals, construction equipment, etc. The economic data shows that rapid recovery or growth of the industrial value added, the GDP, and urban employment all started in the second quarter of 2009, about one or two months after the surge of fixed-assets investment (see Tables 1.9–1.11). This timeline clearly shows the effectiveness of China's stimulus package in driving its economic recovery.

The growth rate of industrial value added accelerated from 4.8 percent in January to 18.5 percent in December 2009; the formal employment in urban areas from – 1.96 million in the first quarter to 1.23 million in the third quarter of 2009. The growth rate of the GDP picked up from 6.1 percent in the first quarter to the 10.7 percent in the fourth quarter of 2009; the whole year growth rate compared to the previous year reached 8.7 percent. According to Ma Jiantang, Director of the National Bureau of Statistics, the contribution of investment to the GDP growth rate was 92.3 percent in 2009.[24]

Thanks partly to Chinese government policies of encouraging exports and partly to gradual recovery of the world economy, China's exports stopped the trend of decline and started recovery from December 2009; its monthly exports have exceeded the highest record before the global financial crisis in January 2010, and they have realized consecutive and rapid monthly growth since then. The realized foreign direct investment started to pick up in August 2009, and has continued to grow since then.

Why has China's anti-crisis macro-economic management been effective?

China's anti-crisis macro-economic policy measures have, by and large, been effective in terms of stopping the decline of manufacturing industries and slowdown of the GDP. Among the G-20 group member countries, China is one of six countries that managed to avoid negative GDP growth in 2009; among the "Positive Six", China's growth rate is the highest.[25]

If we make some comparisons between China's experience of dealing with the current global financial crisis and in the Asian Financial Crisis of 1997–98, we can find some common features in the two cases. On both occasions, the government funded fixed-assets investment on a large scale, which played an important role in offsetting the downward pressures of decreasing external demands on the domestic economy. The Asian Financial Crisis started in July 1997, when China was in the fourth year of implementation of an austere fiscal policy and contractive monetary policy in order to cool down its overheated economy. Chinese leaders were not aware of the impending external shocks until the summer of 1998 when China witnessed a rapid drop in exports, as well as sudden slowdown of investment and consumption. As a result, the GDP growth rate decreased at the speed of 1.8 percentage points in the first six months of 1998.

The Chinese government quickly adjusted the fiscal plan for the year, and decided to expand fiscal expenditure to promote both investment and consumption. It issued 100 billion yuan in long-term government bonds and 270 billion yuan in special government bonds to fund public investment in construction of highways and the Qinghai–Tibet railway, and also increased the salaries of the civil servants to promote private consumption. Some research finds that the proactive fiscal policy contributed 1.5, 2, 1.7, and 1.8 percentage points to China's GDP growth in the period 1998–2001.[26]

However, the expansionary monetary policy implemented in the summer of 1998 was not as effective as it was during 2008–09. Although the central government helped the state-owned banks with their recapitalization in 1998, the banks were still very reluctant to provide loans because of external risks. This made the fight against the Asian Financial Crisis rely solely on central fiscal expansion. In 2008–09, the reactions from banks to government expansionary policy were much more positive. An increase in bank loans actively facilitated the needs of the local government and enterprises for funds to match the construction projects that are partly funded by the central budget and partly funded by loans borrowed by local government and business. The banks' positive attitude toward the expansionary monetary policy is reflected by the fact that the increase in bank loans was as high as 34 percent in 2009, much higher than growth rate the money supply. This greatly improved the effectiveness of the anti-crisis macro-economic policies as a whole in China.

Both proactive fiscal policy and expansionary monetary policies are conventional measures for governments to fight against external shocks or economic crises. Many other governments in the world also implemented similar fiscal and monetary policies during both the period of the Asian Financial Crisis and of the global financial crisis. Some policies implemented by other countries were even more intensive and extensive than that of China. Why have China's anti-crisis macro-economic management measures been more effective than other countries? Are there any advantageous conditions necessary for the effectiveness of the anti-crisis policies that China has but other countries do not have? Brief international comparisons on the matter reveal that China's relatively good fiscal positions, limited exposures to international financial market of China's key financial institutions, and strong state capacities in daily economic activity management are important factors for effective implementation of anti-crisis macro-economic policies.

First of all, effective anti-crisis macro-economic management requires the governments to be in a good fiscal position, thus having plenty of room to make decisive, bold, and timely fiscal expansion to fight against challenges from the demand side. Over the past decade, China's fiscal revenue has been growing at a much faster speed than that of the GDP, and state budget deficit ran relatively low according to international standards. Before the global financial crisis, China ran a small budget surplus of 0.6 percent and a very moderate budget deficit of 0.4 per cent of GDP in 2007 and 2008, respectively (see Table 1.16). To fund a stimulus package, the central government expanded the budget deficit.

Table 1.16 Growth of fiscal revenue and ratio of budget deficit/GDP

Year	*Growth*	*Deficit/GDP*	*Year*	*Growth*	*Deficit/GDP*
1990	10.21	−0.78	2000	17.05	−2.51
1991	7.23	−1.09	2001	22.33	−2.29
1992	10.60	−0.96	2002	15.36	−2.62
1993	24.85	−0.83	2003	14.87	−2.16
1994	19.99	−1.19	2004	21.60	−1.31
1995	19.63	−0.96	2005	19.90	−1.24
1996	18.68	−0.74	2006	22.47	−0.78
1997	16.78	−0.74	2007	32.40	0.60
1998	14.16	−1.09	2008	19.50	−0.42
1999	15.88	−1.94			

Source: Calculated by the author based on the data in National Bureau of Statistics, *Guojia tongji nianjian 2009* [*National Statistical Yearbook 2009*].

But according to Xie Xuren, China's Minister of Finance, the ratio of budget deficit to GDP was under 3 percent in 2010. The ratio of total government debt to GDP was also very low in 2010, only 20 percent.[27] In contrast to other major economies, this relatively good fiscal position gave the Chinese government good financial strength and a flexible way to use proactive fiscal policy effectively to supplement the lack of demand caused by the fall in export demand, as well as the fall in consumption and non-governmental investment demand.[28]

Second, the financial institutions must be in a relatively good shape and function well to facilitate society's needs in order to implement an anti-crisis macroeconomic management policy. There were some important factors that kept China's financial institution and commercial banks in particular in relatively good shape before China was hit by the global financial crisis.

China's financial sector has been very different from those in the United States and Europe before and during the global financial crisis. It has been under the direct administrative control of the state, which produces twofold consequences: while it is inefficient in allocating social capital, it avoids broad exposure to the huge uncertainty caused by unregulated derivatives.

In addition to this, China had just completed overhauls of its banking system after the Asian Financial Crisis. The Chinese government wrote off a large amount of non-performing loans from the balance sheets of the state-owned banks, also injected sizable capital to keep them in a good shape. Because of these actions, China's banking system was relatively safe when the global financial crisis hit China. According to Yongding Yu, former member of Monetary Policy Committee of the PBOC:

> There was no liquidity shortage, no credit crunch, and the monetary multiplier in China has not fallen as dramatically as in the United States. Therefore, the dramatic increase in liquidity in the inter-bank money market has been duly translated into a rapid increase in bank credit and broad money.[29]

In terms of international exposure, China's banks have been very cautious. Although all the major Chinese banks are massive in terms of size, their business reach remains relatively limited in functional and geographical diversification. The world saw the first wave of overseas investments by Chinese banks in 2007, many of which were bolstered by proceeds from initial public offerings, and by rising profits and surging share prices. Prominent examples included China Development Bank's purchase of a 3.2 percent stake in Barclays, China Minsheng Bank's acquisition of a 9.9 percent stake in San Francisco-based United Commercial Bank Holdings, and the Industrial and Commercial Bank of China's purchase of a 20 percent stake in South Africa's Standard Bank. Chinese sovereign entities were also active in investing overseas – the China Investment Corp. made early investments in Morgan Stanley and Blackstone Group, acquiring 9.9 percent of each company.[30] Because of this late participation in the international financial market, China's banks and other non-banking financial institutions were not exposed massively to the subprime mortgage either directly or indirectly through a huge holding of bad assets containing the subprime mortgage. The fact that China's banking system was basically unharmed by the global financial crisis not only avoided a dramatic loss of public confidence in the financial system in China during the crisis, but also allowed it to play important roles in facilitating China's anti-crisis stimulus plan.

Strong state capacities for implementation of macro-economic policies are also a significant factor for the effectiveness of anti-crisis policies in China. The key mechanism through which the government implemented anti-crisis policies was daily involvement of various levels of governments in business affairs and the state-controlled banks and enterprises. After three decades of market-oriented economic reform, the Chinese government still has a strong role to play in organizing economic activities. In many areas, the government is still a dominant force over other players in the market. China is one of a few large economies where the majority of financial institutions and financial assets are directly controlled by the state; so are the land and other key economic resources. State-owned enterprises monopolized important industries, including infrastructure, transportation, telecommunication, and energy. This tight control by the state may cause a huge efficiency loss when the economy runs on good days, but it would quickly avoid the worst result when the economy encounters a crisis. Of course there is a tradeoff between pursuing economic efficiency in the long term and avoiding economic collapse in urgent situations, especially when this crisis came externally.

Challenges of post-crisis macro-economic management

Although there is a debate among those who predict V-shaped recovery and those who predict W-shaped recovery of the world economy, the growth rate of China's GDP is very likely to reach 10 percent in 2011 and to stay around this level for a few years to come. China's exports will also grow at a even faster pace than that of its GDP. Ironically, 10 percent GDP growth was regarded as a symptom of economic overheating some years ago, but it seems that China now feels very comfortable or even feels obliged to keep this growth rate after the global financial crisis.

The truth is that some favorable conditions that are very helpful for China to implement anti-crisis macro-economic policies effectively might be difficulties for its non-crisis-period macro-economic management. And there are always some tradeoffs any government has to face between achieving short-term goals of controlling or offsetting damage caused by urgent crises and realizing long-term goals of balanced, efficient, and sustainable economic development. It's time for Chinese policymakers to consider how to overcome negative consequences brought about by short-term-aimed anti-crisis policies and make a smooth transition from crisis-control management to long-term management.

The first negative consequence of the anti-crisis management or the stimulus package is that the emergent crisis management interrupted China's efforts of economic restructuring. For many years, China's economic growth has appeared to be a clear pattern of heavy dependency on the growth of investment and exports. The shares of capital formation and exports in total GDP have been constantly high since 2001 (see Tables 1.17 and 1.18, Figure 1.5). The external shock caused by the global financial crisis actually reminded us once again how vulnerable this growth pattern is.

A constant high investment rate and lack of domestic consumption will only cause even higher investment rate and more dependence on the international market. This will inevitably cause welfare loss for consumers, overcapacity of production, and inefficient resource allocation. The Chinese government realized the problems of such an economic structure and attempted necessary but painful adjustments before the global financial crisis. However, in order to offset the negative impacts of export decline on the GDP growth and employment, China stopped its efforts and increased fixed-assets investment to keep the economic growth and maintain employment. The stimulus package did not change this unsound growth pattern, instead it made Chinese economy even more addicted to investment.

Another side-effect brought about by the stimulus package is potentially bad loans by China's commercial banks. Among the four-trillion expenditure plan, the majority of funds were made up of bank loans, either lent to the state-owned firms or to the local governments. For the local governments, there is no significant difference between central government direct investment and the loans borrowed from the state-owned or state-controlled banks, they are all free money. That is why provincial and subprovincial-level governments are very enthusiastic to engage in construction projects. In order to match the funds granted by the central government, the investment funds pledged by provincial government were as high as 27 trillion yuan. Without doubt, the enthusiastic local governments promoted a quick investment and increased the current year's GDP growth during the crisis; however, the fast reaction of local government to the economic crisis and their hasty and uneconomical investment decision could also harmful to long-term economic health. In cases where local governments cannot repay the loans to the banks, bad loans will occur and the central government's contingent liabilities may shoot up in the future.

Table 1.17 Shares of investment, consumption, and net exports in GDP

Year	*Investment*	*Consumption*	*Net exports*[a]
1978	38.2	62.1	−0.32
1979	36.1	64.4	−0.49
1980	34.8	65.5	−0.32
1981	32.5	67.1	0.34
1982	31.9	66.5	1.63
1983	32.8	66.4	0.82
1984	34.2	65.8	0.02
1985	38.1	66.0	−4.04
1986	37.5	64.9	−2.43
1987	36.3	63.6	0.09
1988	37.0	63.9	−0.98
1989	36.6	64.5	−1.07
1990	34.9	62.5	2.64
1991	34.8	62.4	2.74
1992	36.6	62.4	1.00
1993	42.6	59.3	−1.84
1994	40.5	58.2	1.26
1995	40.3	58.1	1.58
1996	38.8	59.2	1.97
1997	36.7	59	4.35
1998	36.2	59.6	4.19
1999	36.2	61.1	2.78
2000	35.3	62.3	2.42
2001	36.5	61.4	2.13
2002	37.9	59.6	2.57
2003	41.0	56.8	2.19
2004	43.2	54.3	2.54
2005	42.7	51.8	5.42
2006	42.6	49.9	7.51
2007	42.2	49.0	8.89
2008	43.5	48.6	7.87

Source: National Bureau of Statistics, *Guojia tongji nianjian 2009*.

Note

a Net export share was calculated by the author based on data in Table 1.6.

Although the expansionary monetary policy served the purpose of stimulating economy during economic difficulties, it caused an unnaturally low interest rate, which in turn brought about a huge loss of efficiency of capital allocation. Those who have access to bank loans are usually local governments and state-owned firms, but they are the people who care least about investment efficiency. On the other hand, the private sectors in China are usually more efficient but they have less or no access to the bank loans when the interest rate is very low. This unreasonable asymmetric access of state sector and private sector to the bank loans is caused by the unnaturally low interest rate, which is a consequence of loose monetary policy.

Table 1.18 Contributions of investment, consumption, and net exports to GDP growth

Year	*Investment*	*Consumption*	*Net exports*
1991	24.3	65.1	10.6
1992	34.2	72.5	−6.8
1993	78.6	59.5	−38.1
1994	43.8	30.2	26.0
1995	55.0	44.7	0.3
1996	34.3	60.1	5.6
1997	18.6	37.0	44.4
1998	26.4	57.1	16.5
1999	23.7	74.7	1.6
2000	22.4	65.1	12.5
2001	50.1	50.0	−0.1
2002	48.8	43.6	7.6
2003	63.7	35.3	1.0
2004	55.3	38.7	6.0
2005	37.7	38.2	24.1
2006	42.0	38.7	19.3
2007	39.7	40.6	19.7
2008	45.1	45.7	9.2
2009	92.3	52.5	44.8

Source: National Bureau of Statistics, *Guojia tongji nianjian 2009.*

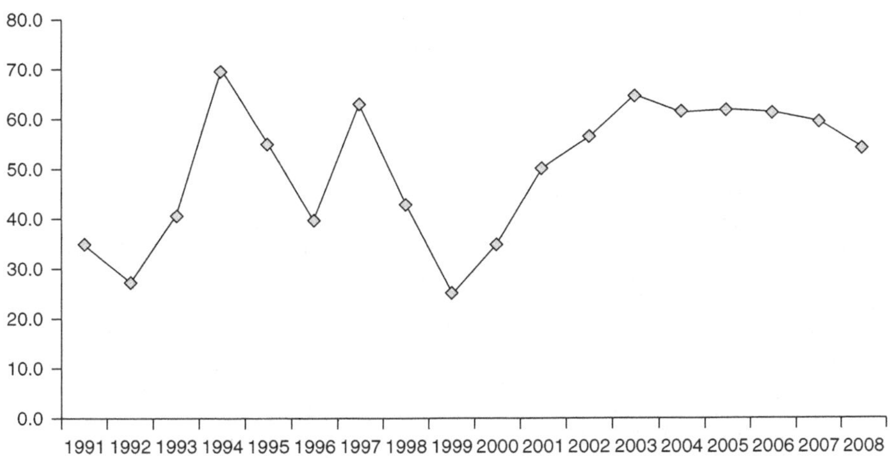

Figure 1.5 Contributions of investment and net exports to GDP growth.

Source: National Bureau of Statistics, *Guojia tongji nianjian 2009* [*China Statistical Yearbook 2009*] (Beijing: Zhongguo tongji chubanshe, 2010).

The final issue China will face is its role in balancing the global economy. China has become the largest exporter and the second largest importer in the world. As mentioned above, due to the considerations of unemployment pressures caused by a sudden decline of international demand for China's products, there was a retreat in China's foreign trade policy. It stopped the efforts of encouraging trade restructuring, and resumed a large amount of export tax rebate to support low-tech and labor-intensive processing industries. As a temporary policy measure to prevent a large surge of unemployment, it is understandable. But how to phase-out such a policy after economic recovery and continue exports and economic restructuring should be a strategic issue for the Chinese government to consider in its post-crisis macro-economic management.

Concluding remarks

The impressive economic performance of China during the global financial crisis demonstrates the effectiveness of China's short-term anti-crisis macro-economic management. While some conventional macro-economic management measures, namely the proactive fiscal policy and expansionary monetary policy, have played important roles in offsetting the negative impacts caused by the external shocks, some special conditions and mechanisms are also key factors that enable state anti-crisis policy to be implemented swiftly. Those conditions and mechanisms include relatively good fiscal position, limited exposure of domestic financial institutions to the troublesome internal financial market, and various levels of government involvement not only in the macro-economic management but also in the daily economic activities through state-owned/-controlled banks and state-owned enterprises. However, anti-crisis management in the short term might conflict with the long-term macro-economic management; this conflict might worsen China's economic structure and damage its efforts of realizing the balanced economic growth.

Notes

1 Mark J. Gasiorowski, "Economic Crisis and Political Regime Change: An Event History Analysis," *The American Political Science Review*, Vol. 89, No. 4 (Dec. 1995), pp. 882–897; and Edgardo Zablotsky, "Political Stability and Economic Growth: A Two Way Relation," *CEMA Working Papers*, No. 109 (1996).
2 Dominique Strauss-Kahn, "Message from the Managing Director and Chair of the Executive Board," in *International Annual Report 2009: Fighting the Global Crises* (Washington: IMF, 2010), pp. 4–5.
3 See International Monetary Fund, *World Economic Outlook Update: An Update of the Key WEO Projections* (Washington: IMF, January 26, 2010), p. 2.
4 According to the WTO, the changes in world exports and imports in 2009 were −23 and −24 percent, respectively, in value (nominal) terms and −12.2 and −12.9 percent, respectively, in volume (real) terms, while changes in China's exports and imports were −16 and −11 percent in value terms and −11.1 and +2.8 in volume terms. Trade in current US dollar terms dropped further than trade in volume terms thanks in large part to falling prices of oil and other primary commodities. See World Trade Organization, *World Trade 2009, Prospects for 2010* (WTO, 2010), p. 10.
5 World Trade Organization, *World Trade 2009, Prospects for 2010*, p. 10.
6 China's holdings of US treasury securities once reached 938.3 billion US dollars in September and October 2009. It replaced Japan as the No. 1 US treasury holder at the beginning of 2009. United States Department of the Treasury, "Major Foreign

Holders of Treasury Securities," announced by US Department of the Treasury and Federal Reserve Board, May, 2010.

7 At the annual spring meeting of the World Bank and International Monetary Fund on April 25, the World Bank raised China's voting rights from 2.77 per cent to 4.42 per cent, which makes it the biggest beneficiary of reforms. The changes shift voting rights 3.13 percentage points from developed countries to developing countries. Increase in China's voting power accounts for 52.7 percent of the increased voting power among developing countries. See World Bank, "World Bank Reforms Voting Power, Gets $86 Billion Boost," *World Bank Press Release*, No. 2010/363/EXT (April 25, 2010).

8 China's export/GDP and import/GDP ratios in 2008 were 33 and 25 percent, respectively, according to the author's calculation based on the data in "Statistical Communiqué of the People's Republic of China on the 2008 National Economic and Social Development," published by the National Bureau of Statistics of China on February 26, 2009.

9 Zhang Wei, "Accessing China's Power Reflected in Foreign Trade," in Lee, Kim, and Woo (eds), *Power and Sustainability of the China State* (London: Routledge, 2009), pp. 119–145.

10 The World Trade Organization, *World Trade 2009, Prospects for 2010*, p. 7.

11 Liu Ligang, "Impact of the Global Financial Crisis on China: Empirical Evidence and Policy Implications," *China & World Economy*, Vol. 17, No. 6 (2009), pp. 1–23.

12 According to *Top 200 in China's Foreign Trade* 2009, an annual report published by *China Customs Magazine*, among the top 200 firms, 153 are foreign-invested firms; their exports account for 77.8 percent of the total exports of the top 200.

13 The correlation coefficient between the changes in growth rates of export and industrial value added is 0.375 and statistically significant at the 0.01 level.

14 Yongding Yu, "China's Policy Responses to the Global Financial Crisis," in *Richard Snape Lecture, November 25, 2009* (Melbourne: Productivity Commission, 2009).

15 There are various estimates on employment involved in export-related industries. According to empirical research conducted by Liu Yong, commodity and service exports provided more than 89 million jobs in 2005. See Liu Yong, "A Effective Solution to the Unemployment Problems," 2005, on the website of State Council Research Centre for Economic and Technological Development.

16 At a press conference held by the State Council's Information Office, the Director of the Rural Working Group in the Central Committee of the Chinese Communist Party told journalists that about 20 million migrant workers lost their jobs in urban areas and had to go back to their home villages. This accounted for about 15 percent of total migrant workers working in Chinese urban areas. See Chen Xiwen, "Speech on China's Rural Situation at the Press Conference" held by the State Council's Information Office in Beijing on February 2, 2009.

17 NDRC (National Development and Reform Commission), "Update on the Development of Four Billion Investment Structure and the Projects Invested by the Central Government," *SDRC (State Development and Reform Commission) News*, May 21, 2009.

18 M1 includes currency, checking account balances (including NOW (negotiable order of withdrawal) accounts and credit union share draft accounts), and travelers' checks. M2 includes M1 plus savings and time deposits, overnight repurchase agreements, and personal balances in money market accounts. Both M1 and M2 (especially M2) are important measures of the domestic money supply.

19 "Statistics of Money Supply" from 2007 to 2009 in *Survey and Statistics* on the website of PBOC (People's Bank of China).

20 NDRC, "Update on the Development of Four Billion Investment Structure and the Projects Invested by the Central Government," *SDRC News*, May 21, 2009.

21 See *Statistics and Data 2010* published by the SAFE (State Administration of Foreign Exchange).

22 "Notice of Further Raising the Tax Rebate for Part of Export Commodities," announced by Ministry of Finance of People's Republic of China, No. 88 (2009) on June 3, 2009.

23 Yao Zhizhong, "800 Billion Yuan Export Tax Rebate Distorts Resource Allocation and Worsens Economic Imbalance," *The 21th Century Economic Herald*, September 22, 2009.

24 See Ding Xiaoqin, "The Final Consumption Contributed 52.5 percent to GDP in 2009," *Xinjing Bao*, February 3, 2009. According to this report, Ma Jiantang said, the contributions of capital formation, consumption including private as well as government consumption, and foreign trade to the GDP in 2009 were 92.3, 52.5, and –44.8 percent respectively.

25 Except for China, other countries which had positive growth rate of the GDP in 2009 are India (6.5 percent), Indonesia (4.5 percent), Australia (1 percent), South Korea (0.2 percent), and Saudi Arabia (0.2 percent) (CIA, *The World Factbook*).

26 Jia Kang, "From Proactive to Sound Fiscal Policy: An Improvement to China's Public Finance System," in Wang Mengkui (ed.), *China in the Wake of Asia's Financial Crisis* (New York: Routledge, 2008).

27 Xie Xuren, "Speech on the Press Conference," held by the State Council Information Office, March 6, 2010.

28 According to *The World Factbook* published by the CIA, the ratios of public debts to GDP of the United States, Japan, Germany, France, and the United Kingdom were 39.7, 192.1, 77.2, 79.7, and 68.5, respectively.

29 Yongding Yu, "China's Policy Responses to the Global Financial Crisis."

30 Jing Ulrich, "Global Financial Turmoil: Implications for China," *J.P. Morgan's Hands-On China Series*, September 18, 2008.

2 Managing political crises in China

The case of collective protests

*Jae Ho Chung**

What constitutes a "political crisis" in China? And how is the state dealing with it before and after it takes place? The word "crisis" (危 機) in Chinese consists of two characters: "danger" and "opportunity." During the Maoist era (1949–76), chaos and instability were not necessarily regarded as bad things as they were seen as the sources of dialectic new beginnings in a long series of permanent revolutions. When "being rebellious is deemed glorious" (*zaofan youli*), it was rather difficult to define political crises in operational terms except for the cases where decisions by the supreme leader designated certain situations as such. How the Red Guard Movement ended in a tragic bloodbath in 1968–69, and with those youths who took part in the movement separated from their families to work in the countryside, is illustrative of these eerie dynamics.[1] Mao Zedong, a historian himself, was fully aware of the ultimate limit with which he was able to push the Chinese populace, a "sheet of sand."

The adoption of systemic reforms and omnidirectional opening as national priorities changed much of the Maoist view of "crisis." As Deng Xiaoping's famous dictum that "China cannot afford chaos" (*Zhongguo bu neng ruan*) well demonstrates, disorder and instability are now widely viewed as serious impediments to national development. Having officially designated the current era as a "strategic opportunity period" (*zhanlue jiyuqi*) for turning China into a great power, if not a hegemonic contender, the Chinese government has repeatedly underlined the imperative of securing stable environments for successful reform and development.[2] That is precisely why Deng Xiaoping's other famous dictum that "stability overrides everything else" (*wending yadao yiqie*) can be seen and heard across China.

The more frequently the tenet of stability gets emphasized in official rhetoric, the more tempted one becomes to ponder about the real extent of instability in China. The worst possible case of instability – i.e., political crisis – points to the collapse of China, which may encompass political and even territorial disintegration of the People's Republic.[3] In the case of the former, the dismantling of the Chinese Communist Party (CCP) may lead to a power vacuum and social disorder; and, in the latter case, the division and secession of territorial and administrative units from the sovereign rule of the People's Republic are sufficient to constitute a political crisis. The temporal immediacy between the Tiananmen tragedy (1989) and the Soviet Union's sudden implosion (1991) led many to believe that China was on the verge of following in its northern communist neighbor's footsteps. Some intuitive similarities notwithstanding, a simplistic comparison of the two states as such offers no persuasive support for the thesis of China's collapse.[4]

Chinese history is replete with precedents of dynastic collapse and territorial disintegration.[5] Historical patterns appear to suggest that, in spite of some temporal variations, the fall of the Imperial Court and/or the territorial disintegration in China was facilitated jointly by the rise of defiant local gentry with independent taxation and conscription powers, the formation of peasant rebels (and secret societies), and foreign military aggression. An obvious question at this juncture would be: is this "triple combination" still valid today?

Given that local gentry became an extinct category and that foreign military aggression is virtually a nonexistent threat to China today, the most fatal ingredient of the three is the emergence of organized collective protests in the cities and countryside against the government authorities. Social instability caused by collective protests is not new to post-1978 China. Even during the Maoist era, protests took place intermittently, though smaller in scale and more spontaneous than organized in nature. Food riots were rampant during the 1950s and 1960s when frequent policy reversals pushed the people to the verge of starvation.[6] When the engineered movement of the Cultural Revolution went completely out of control in its later stage, the Party Center did not hesitate to bring in the People's Liberation Army (PLA) to quell the "enemy of the people," as well as the people, for the sake of safeguarding social stability.

We find a déjà vu situation in the Tiananmen tragedy in that the PLA was once again called on to disband the proponents of democracy and other protestors against the Communist Party. The magnitude of ensuing purges and suppression was such that they virtually uprooted the protectors of democracy activists within the CCP and, consequently, curtailed the democracy movement in China almost to the level of oblivion. While democracy advocates have found a new mission in making voices on human-rights-related issues, today, they hardly pose a threat per se to the communist regime.[7]

Falungong offers yet another crucial case of collective protest that constituted a daunting challenge to the communist regime. As a religious sect equipped with a vertical hierarchy of its own, *Falungong* organized series of anti-government protests on a nationwide basis, including the mass rally in front of the Zhongnanhai (the secluded working and residential district for top Party elites) in 1999. After a few years of tight suppression by Beijing, however, *Falungong* has now largely ceased to be a political threat within China. While we still find intermittent reports on its petty activities here and there in China, the sect is no longer viewed as a grave threat as such, its commercial successes overseas notwithstanding. The Chinese government has proven sufficiently effective in containing the sect within a relatively short period of time.[8]

What transpired in the Autonomous Regions of Tibet and Xinjiang in 2008 and 2009, respectively, provides yet another clue as to what type of major political

Table 2.1 Annual growth rates of disorder indicators, 1994–2004

Indicator	*Growth (%)*
Labor disputes	29.9
Collective protests	17.8
Criminal activities	11.0
Administrative litigations	10.2
Civil suits	6.2
Petitions/letters	5.7

Source: Hu Lianhe et al., *Dangdai zhongguo shehui wending wenti* [*Problems of Social Stability in Contemporary China*] (Beijing: Hongqi chubanshe, 2009), pp. 37, 41, 42, 44, 51, 57.

crisis Beijing may have to face in the future. Unlike the democracy movements and sectarian resistance noted above, ethnic-based protests are perhaps more uncompromising in nature and likely to spiral over time, as well as with crucial implications for China's relations with neighboring states. Yet, such crises fall outside the purview of this study as it is covered in another chapter of this volume.[9]

This study consists of three sections. The first provides an operational definition of collective protest as a potential key source of political crisis in China and, then, gauges the changes in its frequency and magnitude since the early 1990s. The second section examines three dimensions of Beijing's efforts to cope with collective protests: namely, prevention, monitoring, and containment. The third section offers an assessment of China's state capacity in averting political crises in the future.

Collective protests as a source of political crises

A survey conducted in 2004 by the Chinese Academy of Social Sciences revealed that 73 percent of the respondents singled out rural/peasant problems as the ultimate source of instability in China's future.[10] According to another survey conducted in 2004 on 107 senior officials studying at the Central Party School, the top four problem areas selected were: (1) income disparities (chosen by 44 percent of the respondents); (2) social instability (24 percent); (3) corruption (8 percent); and (4) unemployment (4 percent). As income disparities, corruption, and unemployment are key factors that have affected the level of social instability in China, nearly 60 percent of the respondents also replied that maintaining stability was the most important task of all.[11]

If we look at six different indicators of social disorder in China, it becomes obvious as to why collective protests are deemed the most important source of threat to social stability. As Table 2.1 illustrates, labor disputes and collective protests recorded the highest rates of growth during the ten-year period (1994–2004) concerned. The occurrence of labor disputes is generally limited to the working class in urban areas and, more often than not, they take place in confined premises. On the other hand, collective protests occur in both the countryside and cities and, therefore, collective protests may pose potentially a more serious threat to the regime in power. More importantly, labor disputes that were not resolved within the premises often pushed workers on to the streets, demonstrating and protesting against government authorities.[12]

Defining and gauging collective protests

Collective protests refer to "collective public security incidents" (*quntixing zhian shijian*: widely abbreviated as *quntixing shijian*).[13] According to the most popular definition, collective public security incidents refer to the "incidents whereby a group of 5 or more people illegally gather to disrupt public order and destroy public safety and public properties" (*feifa juzhong souruan shehui zhixu puohuai shehui gonggong anquan yu gonggong caichan de shijian*). Collective protests generally denote incidents that entail such actions as parading illegally, surrounding government offices, attacking government officials, disrupting public transportations, fasting and committing suicide in public, and so on.[14] This definition does not touch upon motives or intentions per se but, given that collective protests are generally distinguished from ordinary crimes, simple labor disputes, and individual petitions, they generally entail offenses of a more organized and political nature.[15]

According to the Central Commission on the Comprehensive Management of Public Security (*zhongyang shehui zhian zonghe zhili weiyuanhui*), as of 2003, collective protests occurred in 257 (99.2 percent) of 259 prefecture-level cities and in 792 counties (53 percent of all counties) in China.[16] Despite their growing importance, however, much of the pertinent data and key information is lacking.[17] The daunting problem of missing data perhaps reflects the serious nature of the situation where collective protests have become prevalent and destabilizing, as well as the preference of the central government for containing all such sensitive data.

Table 2.2 provides a longitudinal trend of collective protests in the whole of China during the period 1993–2008.[18] Despite a few missing data points, it offers a couple of useful observations as to the frequency and magnitude of collective protests since the early 1990s. First, over this sixteen-year period, the occurrence of collective protests increased 14.3 times. As the annual rate of increase standing at 21.6 percent for 1993–2005 clearly demonstrates, the frequency of collective protests rose dramatically, without a single year witnessing a decline.[19]

Some Chinese writings contend that the overall frequency of collective protests has been "stabilizing" since 2006.[20] Other sources suggest without presenting concrete figures that the frequency of collective protests actually increased during 2007–9. One source reveals that the total number of collective protests in Guangzhou during the first half of 2009 increased by 43.2 percent compared to that of 2008. Yu Jianrong, an outspoken expert on the theme, commented in his speech to the Beijing Lawyers' Association in February 2010 that the number of collective protests during 2006–9 "exceeded 90,000 each year" (i.e., surpassed the figure of 2005). While he was not able to give precise figures for obvious reasons, the upward trend was nevertheless confirmed.[21]

Second, the number of participants in the collective protests also rose every year for the same period. The annual average rate of increase stood at 17.8

Table 2.2 Number of collective protests, 1993–2008

Year	*Frequency*	*Participants (1,000 people)*
1993	8,909	N/A
1994	10,000	730
1995	11,000	860
1996	12,000	1,010
1997	17,000	1,180
1998	25,000	1,390
1999	32,000	1,630
2000	37,000	1,910
2001	43,000	2,240
2002	50,400	2,620
2003	58,000	3,070
2004	74,000	3,760
2005	87,000	N/A
2006	90,000 + [a]	N/A
2007	N/A	N/A
2008	127,467	12,176

Sources: Chen Jinsheng, *Quntixing shijian yanjiu baogao* [*A Research Report on Collective Incidents*] (Beijing: Qunzhong chubanshe, 2004), pp. 62, 170; Murray Scot Tanner, "China Rethinks Unrest," *The Washington Quarterly*, No. 3 (Summer 2004), p. 139; Robert Marquand, "In China, Stresses Spill Over into Riots," *Christian Science Monitor*, November 22, 2004; Richard McGregor, "Data Show Social Unrest on the Rise in China," *Financial Times*, January 19, 2006; Jiang Xun, "Zhonggong miandui guanzhi weiji" [Communist China Faces a Governance Crisis], *Yazhou zhoukan* [*Asia Weekly*], December 23, 2007, p. 32; Mu Muying, "Zhou Yongkang chengren guanbi minfan" [Zhou Yongkang Admits that Popular Protests Were Caused by Officials], *Zhengming-Dongxiang* [*Contend-Trend Magazine*], February 2009, p. 10; and Hu Lianhe et al., *Dangdai zhongguo shehui wending wenti* [*Problems of Social Stability in Contemporary China*] (Beijing: Hongqi chubanshe, 2009), p. 60.

Note

a Data cited in Yu Jianrong's presentation at the 6th Beijing Forum held at Peking University on November 7, 2009.

percent for 1994–2004. Unlike the rate of increase in the frequency that varied from the lowest of 9.1 percent (1995–6) to the highest of 66.7 percent (1998–9), the rate of increase for the number of participants stayed within the range of 16.8 to 22.5 percent. In fact, the average number of participants per protest was 74 in 1994, 83 in 1996, 56 in 1998, 52 in 2000, 52 in 2002, and 51 in 2004. If the data for 2008 given in Table 2.2 is reliable, the number of participants per protest was 96, showing a dramatic increase for that year.[22]

Third, the rising frequency and expanding size of collective protests had much to do with the fact that the occupational base of protesters has been widening in recent years. According to a reliable source for 2000, 50 percent of the protest participants came from the countryside, and city dwellers made up 34 percent, while the remaining 16 percent came from religious organizations and ethnic minorities.[23] With the frequency of collective protests continuously rising, all walks of life including intellectuals, private entrepreneurs, taxi drivers, PLA soldiers, veterans, and even civil servants took part in protesting.[24]

Fourth, if the frequency data for 2008 is credible, that means an average of 11.3 protests took place everyday in each of the 31 province-level units in China. Is that a big deal for the CCP? An answer would depend on what kind of protests they were. According to a recent study, the number of large-scale collective protests involving more than 500 people has been rising steeply. While the average annual frequency of such large-scale collective protests was 15.8 for 2003–6, that for 2007–9 rose to 61.7.[25]

Fifth, some national-level data suggest that the level of violence in collective protests has also been rising. In 2003 alone, over 3,100 incidents of blocking public roads and railways took place.[26] Some local statistics also point to the increased level of violence. In Jilin, the number of collective protests entailing attacks on public transportation facilities and government buildings and personnel increased 2.3 times in 2001 compared to 2000.[27] Direct attacks on the police have also become quite common. In Henan, for instance, the injuries and deaths of police officers caused by collective protests accounted for 25 percent of the total injuries and deaths.[28] According to some 2008 data, 1,120 police officers and government officials were killed or injured in collective protests.[29]

Three psychological factors have jointly operated in enhancing the level of violence: (1) as Zhu Rongji himself admitted in 2001, up to 80 percent of the protestors' complaints were well-grounded and could have been reasonably accommodated,[30] which means that protestors who regarded their behavior as "rightful resistance" tended to think that only rough actions would result in government attention and immediate solution; (2) popular beliefs at the grass-roots that "no trouble, no solution; a small trouble, a small solution; a big trouble, a generous solution" (*bunao bu jiejue xiaonao xiao jiejue danao da jiejue*) have also driven protestors to rely on tougher and more violent actions;[31] (3) for many local officials, suppressing protestors "as fast as possible" (*yi kuai zhi kuai*) was deemed politically advantageous to their careers and, therefore, security forces were deployed at will, often prematurely, thereby inviting fierce resistance and, consequently, increasing the overall level of violence involved.[32]

Sixth, contrary to popular assumptions, collective protests take place in poor as well as rich regions. A report by the Macro-economic Research Institute under the now-defunct State Planning Commission identified three regions particularly susceptible to collective protests: (1) provinces heavily endowed with state-owned enterprises – most notably, Liaoning, Jilin, and Heilongjiang; (2) old industrial bases located along the "third front" – Hunan, Sichuan, Chongqing, and Hubei; and (3) some provinces with advanced rural economy – Jiangsu and Zhejiang. Conspicuously missing from the list were the middle-belt provinces with a poor rural economy – Henan, Anhui, and Jiangxi.[33] According to a more recent study, such rich provinces as Guangdong, Jiangsu, and Shandong were also listed as high-risk areas for collective protests, along with poor provinces like Henan, Hunan, and Shaanxi.[34] That is, both wealthy and poor regions have their own reasons for large-scale collective protests. Whereas subsistence and survival were the core motives in most backward regions, the search for justice and the quality of life mattered dearly in many of the advanced regions.[35]

The six observations above provide a sufficient ground on which to argue that collective protests have indeed become a critical source of political instability in China. In terms of frequency, magnitude, violence, and regional spread, collective protests have potentials to pose serious threat to the regime in power. As Beijing's threat perceptions intensified, the government authorities have come up with various measures of preventing, monitoring, and containing them. If the past track records are any useful guide, these measures appear to have thus far been quite effective, at least in downsizing them to a threat that the communist regime is capable of handling with confidence.

Managing collective protests: from prevention to containment

From the viewpoint of the Communist Party and the central government, "managing" collective protests entails the following three functional dimensions: (1) the preventive task designed to reduce and mitigate key sources of social discontent on a nationwide basis; (2) the monitoring dimension of collecting intelligence and conducting surveillance at local and grass-roots levels; and (3) the containment phase where, once taking place, popular protests are effectively contained and suppressed by public security forces and, if necessary, by the military.

The preventive stage

In efforts to prevent the spread of social discontent as the root cause of collective protests, the central party and government apparatus have implemented several categories of measures. First, tactics of "buying off" (*huaqian maipingan*) were widely employed in reducing the economic base of collective protests. Since heavy taxation had long been a key source of contention in the countryside, the ban on illegal levies and the nationwide abolition of the agricultural tax in 2005 was one principal measure designed to lighten the burdens of poor peasants. In mitigating regional and income disparities, a wide range of schemes of fiscal transfers and progressive taxation has also been implemented.[36] Furthermore, since nearly two-thirds (65.1 percent) of all rural collective protests were related to or caused by the local government's grabbing of peasant land, more strict rules were devised and applied to farm land appropriation. With the number of rural residents whose farm land was appropriated by the state standing at 40 million, Beijing issued stern warnings against local officials mishandling the task by way of forced evacuation and under-compensation.[37]

Second, Beijing attributed the rise of collective protests in part to the inefficacy of the system of petitioning. That is, if the petition system as the first channel of detecting discontent and airing grievances had functioned well, many of the collective protests could have been avoided. At least three pieces of data are supportive of the failure of the petition system: (1) as high as 30 percent of the collective petitioners paid repeated visits to make their claims heard; (2) the percentage of collective petitioners in total petitioners rose from 59.8 in 1998 to

75.5 in 2001; and (3) the increase in the number of people bringing their cases to Beijing was higher (14 percent) than that to the provincial capital (0.1 percent) or to the county seat (– 2.4 percent). Naturally, Beijing has sought to improve the nationwide petition system. As a result, the "Revised Regulations for the Petition System" (*xiuding xinfang tiaoli*) were announced in January 2005, which made local officials more sensitive and responsive to petitioners' claims and provided more protection for petitioners.[38]

Third, the central government has also resorted to a wide range of legal measures in coping with collective protests. With the number and scale of collective protests growing, Beijing enacted laws and regulations against collective protests. To list a few, on August 28, 2005, a "law governing public security in the People's Republic of China" (*zhonghua renmin gongheguo zhian guanli chufa fa*) was passed. In September 2006, a "statute on regulating mass activities" (*quntixing huodong guanli tiaoli*) was issued, which requires the organizers of an activity that involves more than 1,000 people to apply for an official approval from public security departments 20 days in advance. Violators are to be charged with fines of 100,000 to 300,000 yuan. On August 30, 2007, the National People's Congress passed a new "law on coping with accident incidents" (*zhonghua renmin gongheguo tufa shijian yingduifa*).[39]

Fourth, certain political measures were implemented to reduce government errors as a source of popular discontent and collective protest. For instance, Chen Jiping, head of the Administrative Office of the Central Commission for the Comprehensive Management of Public Security, acknowledged that local government errors were deemed a primary cause for collective protests in China.[40] By defining the mistakes and misconduct on the part of local governments as the key source of collective protests, Beijing sought to pass the buck and carve out some space for its own maneuvering. At the same time, it stressed a leadership responsibility system (*lingdao zerenzhi*) by which local leaders were held responsible for what transpired in their jurisdictions (*shui guanxia shui fuze*). In Guangdong, for instance, local officials were told that they would be immediately fired if a large-scale collective protest occurred in the areas under their rule.[41]

Fifth, the central party and government of late adjusted its view on collective protests. For one, Beijing came to realize that maintaining "absolute stability" (*juedui wending* or *gangxing wending*) – i.e., zero protest – was an impossible goal to achieve.[42] For another, the center began to see some value in allowing peaceful protests as a useful channel of letting public discontents out, thereby reducing the base for more violent and better-organized anti-government protests. It was in this context that, since 2007, the public security organs became more a bit more tolerant of peaceful parades and mass activities.[43]

The monitoring dimension

Despite the government's preventive efforts noted above, a certain discontent is bound to grow and will generate collective protests. Therefore, extensive and intensive monitoring constitutes a key priority in Beijing's task of coping with them. Two dimensions are particularly pertinent in this regard: (1) the establishment of nationwide surveillance and intelligence networks in both rural and urban areas; and (2) the deployment of Internet control mechanisms.

For the establishment of nationwide surveillance and intelligence networks, both central and local hierarchies have been in operation. At the central level, public security-related ministries and agencies – i.e., Ministry of Public Security, Ministry of National Security, State Council Information Office, and so on – engage in gathering intelligence and monitoring local situations. These units report directly to the State Council and Premier. They also operate in close coordination with at least three units on the Party side: the Central Commission for Comprehensive Management of Public Security designed to deal specifically with the problem of collective protests, the "Stability Preservation Leading Group" (*weihu wending lingdao xiaozu*) led by Zhou Yongkang (Politburo Standing Committee member) with a suboffice called the "Stability Preservation Office" (*weiwenban*), and the Central Commission on Political and Legal Affairs (*Zhongyang zhengfa weiyuanhui*) with broader missions.[44]

While the precise nature of the division of labor among these units remains veiled, the Stability Preservation Leading Group is the nucleus in terms of managing collective protests.[45] Serious cases of collective protests are discussed and solutions are sought in meetings of this unit. Very serious cases may even be sent to the Politburo Standing Committee for deliberation, where Hu Jintao chairs and Li Changchun and Zhou Yongkang are members. Violent large-scale protests in Tibet and Xinjiang in 2008 and 2009, respectively, met those criteria.[46]

As for the street-level surveillance, a distinction needs to be made with regard to the rural/urban divide. Generally speaking, from a public-security management perspective, cities with confined spaces – though more densely populated – are relatively easier to contain than the vast countryside. Whereas the task of monitoring in the cities is carried out by roadside patrol, twenty-four-hour surveillance, and rapid-reaction forces (e.g., SWAT), the same cannot be said of the countryside with expansive areas and fewer police officers. In Hunan, as of 2006, the number of police stations with less than five officers amounted to 549 and there were 418 stations without a single police vehicle, all located in the rural area.[47]

In efforts to strengthen the capacity for monitoring rural areas, Beijing implemented three measures. First, from 2002 on, the concept of "community police control" (*shequ jingwu*) was promoted, according to which at least one police officer was assigned to monitor every 1,000 households (or 3,000 people) in the countryside. By 2007, a total of 220 thousand police officers were assigned to more than 143 thousand "police offices" (*jingwushi*). At least 83 percent of China's vast rural areas now have the presence of at least one police officer.[48] Second, in order to mitigate the shortage of police personnel in the rural area, Beijing reinvigorated the "rural security maintenance association" (*nongcun zhian baoweihui*) under the villagers' committee, whose function is stipulated in Article 111 of the Constitution.[49] Third, since 2006, peasants (and urban residents as well) were encouraged to sign up to be "voluntary information officers" (*zhiyuan xinxiguan*) who were required to formally register at the county's or city's stability preservation office.[50]

It was in the process of coping with the followers of *Falungong* in the late 1990s that the Chinese government authorities became keenly aware of the formidable power of the Internet in facilitating anti-government collective protests. Beijing's efforts for Internet control have centered on the following four areas. First, pertinent government units have carried out close monitoring of numerous Internet sites and their contents by way of monopolizing the regulation of Internet service providers (ISPs) and Internet content providers (ICPs). Second, the State Council as well as provincial authorities set up special bureaus and task forces empowered to screen news contents on the Web and e-mail communications. As of 2004, the Chinese government had over thirty-thousand people who routinely scanned e-mails for disruptive and subversive messages.[51]

Third, regulations were also promulgated so that Internet contents and service providers were required to keep records of all site users and their contents for up to sixty days, allowing them to be turned over to the police on demand.[52] Fourth, since nearly 30 percent of large-scale collective incidents in China were found to be related to Web postings and cell-phone communications (SMS) – often dubbed "online collective protests" (*wangshang qunti shijian*) – government authorities devised measures to limit people's right to set up personal websites. In June 2009, for instance, the central government required the manufacturers to install Internet filtering software (*luba*) on all new computers.[53] As online technologies were advancing rapidly, the Chinese government's countermeasures have also been evolving in tandem. In early 2010, for instance, Beijing set up a new office called the Internet News Coordination Bureau (known as Bureau No. 9) to monitor and curb social networking sites, microblogging sites, and video-sharing sites.[54]

The containment mechanism

If the preventive and monitoring mechanisms should all fail and large-scale violent collective protests should cause grave threat to the communist regime, the ultimate solution would lie in the use of physical forces at Beijing's disposal. Two issues are pertinent in such contingencies: (1) the command structure; and (2) the menu for choices in containing the spread of collective protests.

Most of the collective protests in China are handled at the local (mostly subprovincial) level, although their daily occurrences and trends are routinely reported to the Stability Preservation Leading Group and the Central Commission for Comprehensive Management of Public Security. With regard to "ordinary collective protests" (*yibanxing quntixing shijian*) and "somewhat large collective protests" (*jiaoda quntixing shijian*) – which involve five to 100 people and 100 to 300 people, respectively – a two-level command structure encompassing the city/prefecture and county authorities is deployed. As for "very large collective protests" (*zhongda quntixing shijian*), which involve 300 to 1,000 and "extraordinarily large collective protests" (*tebie zhongda quntixing shijian*), which involve more than 1,000 people, the provincial government is put in charge of offering unified leadership. Regarding cases of large-scale protests whose boundary spans multiple provinces, the central authorities may also get involved.[55]

Beijing's menu for choices in dealing with collective protests also varies with their scale and impact. Since most of the collective protests in China took place in geographically limited areas, generally within a county's premise, Beijing found it relatively easy to contain them by way of utilizing a small force of civilian police (*minjing*) under the Ministry of Public Security (MPS).[56] Violent protests, though limited in geographical scope, were often suppressed by the deployment of the People's Armed Police (PAP) under the dual command of the State Council (through the MPS) and the Central Military Commission. According to a source, collective protests involving more than 1,000 people (i.e., extraordinarily large collective protests) are generally dealt with by the PAP, which has at least a division-sized force stationed in each province-level unit.[57]

The PLA constitutes the last resort for maintaining stability and protecting social order in the face of very large and violent collective protests.[58] Article 22 of the National Defense Law passed in 1997 stipulates that "[T]he standing army, when necessary, may assist in maintaining order in accordance with the law."[59] Unlike the earlier (1998, 2000, and 2002) versions, the 2004 edition of China's *Defense White Paper* specifically lists "cracking down on criminal activities of all sorts and maintaining public order and social stability" as one of the PLA's key functions.[60] Furthermore, the PLA is now required to perform the so-called "diverse duties" (*duoyanghua renwu*), which include maintaining social stability, in addition to fighting epidemics, combating natural disasters and contributing to economic development.[61]

There is not only a legal and political basis for the PLA's involvement in dealing with collective protests, the precedents where the military was utilized in the Tiananmen Square in 1989 and in Tibet and Xinjiang in 2008 and 2009, respectively, clearly point to such a choice at Beijing's disposal. Depending on which type of force is deployed in containing collective protests, the command structure may also vary. First, when it requires nonmilitary operations – i.e., the civilian police force is used – provincial or subprovincial party and government leaders are put in charge. Second, when the PAP is deployed and the operation at hand is defined as semimilitary, provincial leaders and military leaders in the provincial military district get to share the leadership. Third, if the PLA should be called on, then the commanders of the provincial military district assume exclusive leadership in accordance with the order from the Central Military Commission.[62]

In tandem with the newly designated "diverse duties" of PLA and the growing need for all-out mobilization to deal with accidental incidents, a system of so-called "national defense mobilization" (*guofang dongyuan*) has received much attention. According to this scheme, civilian and military authorities are required to maintain close coordination in command, and military, semimilitary, and civilian security forces are to be deployed not only for combatant situations but also for nonwar emergencies such as containing large-scale social disorder.[63] In fact, recently, the need for reinvigorating the militia (*minbing*) has been stressed by many military publications.[64]

Assessing the future: state capacity vs. collective protests

In sum, collective protests have not yet reached a stage where a nationwide or interprovincial united front has been formed to wage single-minded attacks against the Communist Party and the central government. Most of the protests are still largely grass-roots and local in their scope and more economic than political in nature, hardly necessitating the deployment of the PLA. Yet, in tandem with the rising frequency, violence levels, and magnitude of collective protests, the state's countermeasures have also been evolving and diversifying.

As in the case of its exceptionally viable strategies for economic reform, we should not underestimate China's learning capabilities in dealing with domestic challenges. The tumultuous years of the Cultural Revolution decade and the sudden implosion of the Soviet Union offered China sobering but extremely useful lessons. Beijing has since the early 1990s sought hard to raise its governance capacities in order to get prepared for all sorts of contingencies, thereby sustaining stability by all means. Strenuous efforts have been made to strengthen Beijing's fiscal control over localities, to tighten its control over ethnic minority regions, to fight against corruption and organized crime, to minimize the room for unorthodox ideas and beliefs, to augment Party control over grass-roots units and clans, and so on.[65]

In the short- to mid-run, the most likely scenario would be the one that is often referred to as "stable instability."[66] This scenario denotes a situation where collective protests do frequently take place and even become almost routines, but they do not necessarily amount to threatening the fate of the communist regime. As it stands now, collective protests are not going to pose a systemic threat to China since it is highly unlikely for these incidents to become a nationwide challenge. While collective protests might take on a significantly different character if some ideologies or religious beliefs should fill in the ideational void created by the demise of Communism, it is nevertheless difficult to foresee a situation where religious movements or ethnic minority riots are organically and laterally linked to collective protests to cause nationwide armed resistance against the central government.[67]

Given that discontent and dissatisfaction do not necessarily or automatically lead to disputes and that not all disputes are systematically led or become violent revolts, the frequency and magnitude of collective protests alone are not sufficient conditions for posing an apocalyptic scenario for China. Furthermore, in many cases, the central government has been viewed by the ordinary people as a potential mediator and even key supporter on the side of protestors against "corrupt" local authorities.[68] Without fundamental political reforms to set up channels through which demands of the populace are freely expressed, articulated, and represented in policymaking, Beijing's governance capacity may stagnate or decline in terms of enhancing responsiveness and legitimacy of the regime.

In sum, the frequency of collective protests may not matter so much unless they constitute single-minded attacks on the regime in Beijing, unless they form a single nationwide anti-regime united front, and unless their size, scope, and violence dictate the deployment of armed forces that go beyond the PAP. Of course, this overall assessment regarding China's capacity of managing political crises stemming from collective protests excludes a situation where the intervention of a crucial catalyst – which no social scientific inquiry can possibly foresee – is made.

Such a catalyst may consist of three components. First, it requires an intricate issue that is both appealing to a nationwide audience and resonates well with the atmosphere of the time, although forecasting its precise nature appears impossible at this stage. Second, the formation of the middle class who are intrinsically different from the pro-CCP entrepreneurs (i.e., "red capitalists") may also introduce a new variable in the state-society dynamics in China. Third, as the communist state has become equipped with increasingly more effective means of control than ancient dynasties, ordinary people today have also come to enjoy more advanced weapons of the weak – the Internet (over 400 million users) and mobile phones (750 million subscribers) – that can reduce the cost of organizing to virtually zero.[69]

So long as no such catalysts come into joint play, stable instability may persist in the short to medium run. Will historical perceptiveness and reflective learning be sufficient to save China from an ill-fated future? To what extent the various countermeasures of the state – that range from canceling the agricultural taxes and tightening the rules of farm land appropriation to enhancing surveillance capacity and centralizing the control of public security forces – will be able to produce desired outcomes remains to be seen. Then, again, collective protests are but one – though important – determinant of China's uncertain future. After all, many great powers before China, including Great Britain, the United States, and Japan, went through similar phases of unrest and disorder in their pursuit of industrialization and telescoped development. Yet, they survived to attain governed stability.

Notes

* I would like to acknowledge useful comments on an earlier version by Hon S. Chan and Paik Wooyeal.

1 See Hong Yung Lee, *The Politics of the Chinese Cultural Revolution: A Case Study* (Berkeley: University of California Press, 1978), ch. 9; and Tony Saich, *Governance and Politics of China* (Basingstoke: Palgrave, 2001), pp. 164–166. For a counter-narrative to the "catastrophe" (*haojie*) discourses on the Cultural Revolution, see Mobo Gao, *The Battle for China's Past: Mao and the Cultural Revolution* (London: Pluto Press, 2008), ch. 1.

2 For this logic, see Li Liangdong and Hou Shaowen, *Wending: yadao yiqie de daju* [*Stability: The Foremost Goal that Overrides Everything Else*] (Beijing: Zhongyang dangxiao chubanshe, 1999); and Chen Peixiao and Xia Liping (eds), *Xinshiji jiyuqi yu zhongguo guoji zhanlue* [*Opportunities of the New Century and China's International Strategies*] (Beijing: Shishi chubanshe, 2004).

3 Principal indicators of a political crisis in China may entail the simultaneous eruption of violent anti-government protests in different regions, the imposition of a nationwide martial law with the involvement of the People's Liberation Army as the fixer, and a considerable loss of control at various local levels, which may happen in conjunction with serious cracks in leadership unity.

4 For a critical analysis of such simple analogies, see Jae Ho Chung, "Assessing the Odds against the Mandate of Heaven: Do the Numbers (on Instability) Really Matter?" in Jae Ho Chung (ed.), *Charting China's Future: Political, Social and International Dimensions* (Lanham: Rowman & Littlefield, 2006), pp. 107–109.

5 According to a study, 44 percent of the 2,132 years from the Qin Dynasty to the demise of the Qing (221 BC– 1911) saw a divided and disintegrated China. See Zhang Xianglin, *Wendinglun* [*On Stability*] (Beijing: Zhongyang wenxian chubanshe, 2004), p. 14.

6 See Cheng Meidong (ed.), *Dangdai zhongguo zhongda tufashijian* [*Major Accidental Events in Contemporary China*] (Beijing: Xinhua shudian, 2008), pp. 36–51.

7 For this line of assessment, see Jean-Philippe Beja and Merle Goldman, "The Impact of the June 4th Massacre on the Pro-Democracy Movement," *China Perspectives*, No. 2 (2009), pp. 18–29.

8 See Jae Ho Chung, "Challenging the State: Falungong and Regulatory Dilemmas in China," in John D. Montgomery and Nathan Glazer (eds), *Sovereignty under Challenge: How Governments Respond* (New Brunswick, NJ: Transaction, 2002), pp. 83–106; Maria Hsia Chang, *Falungong: The End of Days* (New Haven, CT: Yale University Press, 2004); and James W. Tong, *Revenge of the Forbidden City: The Suppression of the Falungong in China, 1999–2005* (Oxford: Oxford University Press, 2009), Chs 6 and 8.

9 Refer to Colin Mackerras's contribution (Chapter 4).

10 George J. Gilboy and Eric Heginbotham, "The Latin-Americanization of China," *Current History*, No. 674 (September 2004), p. 257.

11 *Beijing Ribao* [*Beijing Daily*], November 29, 2004, reported in *Xinhua News Service*, November 29, 2004.

12 In some cases, such merged concepts as "collective labor disputes" (*quntixing laodong zhengyi*) and "collective petitioning" (*quntixing shangfang*) are also used. See Zhu Chao, "Quntixing laodong zhengyi de yufang xiaojie yu jizhi jianshe yanjiu [Study of Preventing and Resolving Collective Labor Disputes and of Institution-Building]," http://www.usc.cuhk.edu.hk/wk_wzdetails.asp?id=3321 and "Neimenggu quntixing shangfang shuliang duo guimo da [Collective Petitioning in Inner Mongolia Increases in Number and Scale]," *Neibu canyue* [*Internal References*], No. 28 (July 26, 2002), pp. 32–34.

13 Oftentimes, the term "accidental incidents" (*tufa shijian*) is used interchangeably with collective protests. But the former is in fact a broader concept as it encompasses natural disasters, industrial accidents, endemics/pandemics, environmental catastrophes, and terrorist attacks, as well as collective protests. See the Writing Team (ed.), *Yingdui tufa shijian* [*Coping with Accidental Incidents*] (Beijing: Xinhua chubanshe, 2008), pp. i, 1. For Chinese glossary that refers to collective protests, see Task Force Team (ed.), *Zhongguo quntixing tufa shijian – chengyin ji duice* [*Collective and Accidental Incidents in China: Reasons and Responses*] (Beijing: Guojia xingzheng xueyuan chubanshe, 2009), p. 2.

14 See Hu Lianhe et al., *Dangdai zhongguo shehui wending wenti* [*Problems of Social Stability in Contemporary China*] (Beijing: Hongqi chubanshe, 2009), pp. 57, 139.

15 Whereas most collective protests were initially triggered by economic, religious, ethnic, justice-seeking, and other reasons, they took on a political character once they became organized, violent and anti-government in nature. See The Task Force Team (ed.), *Zhongguo quntixing tufa shijian*, pp. 6–10.

16 See http://www.renminbao.com/rmb/articles/2004/6/21/3157.html (last accessed on July 7, 2004).

17 See Cheng Tongxun, *Dangdai zhongguo nongcun zhengzhi fazhan yanjiu* [*Study of Rural Political Development in Contemporary China*] (Tianjin: Tianjin renmin chubanshe, 2000), pp. 57–60.

18 The nationwide data on collective protests that are both consistent and openly circulated in China cover only up to 2005. The post-2005 data (of which data on 2006 and 2008 are cited in this table) are unofficial and may have reliability problems.

19 See Cai Xia, "Qinfang guojia yu shehui guanxi nishi fazhan [Prevent State-Society Relations from Being Reversed]," *Gaige neican* [*Internal References for Reform*], No. 25 (2009), p. 3; and Ren Yuling, "Jianshao tufa shijian fasheng shi yingji guanli de zhongyao shiming [Reducing the Occurrence of Accidental Incidents Is the Core Mission of Crisis Management]," *Gaige neican*, No. 34 (2009), p. 34.

20 See, for instance, The Task Force Team, "Jingji jiegou tiaozhengqi quntixing shijian fazhan qushi tezheng ye yufang [The Development, Characteristics and Prevention of Collective Protests during the Period of Economic Structural Adjustment]," *Lilun dongtai* [*Theory Dynamics*], No. 1853 (June 10, 2010), p. 2.

21 Shan Guangding, "2009nian shangbannian quntixing shijian de taishi he tedian [Trends and Characteristics of Collective Protests in the First Half of 2009]," *Lingdao canyue* [*References for Leaders*], No. 28 (October 5, 2009), p. 11; and Yu Jianrong, "Maintaining a Baseline of Social Stability," in *China Digital Times* (February 20, 2010) at http://Chinadigitaltimes.net (accessed on May 6, 2010). For an assessment that the frequency of collective protests has increased, see Xie Wuzhong, "Guofang dongyuan xitong chuzhi quntixing tufa shijian wenti xiantan [On the Issue of Having the National Defense Mobilization System Deal with Collective Protests and Accidental Incidents]," *Guofang* [*National Defense*], No. 1 (2010), p. 24.

22 Given that 2008 was the year of the Olympic Games, two possibilities are conceivable. First, despite the seemingly near-perfect stability in Beijing, more protests were reported elsewhere (Tibet, Sichuan, Gansu, and so on) in China. Second, being acutely aware of the government's extreme sensitivity during the Games, people were perhaps more keen to take advantage of the timing to push for their claims through organized protests. In fact, new modes of collective protest – e.g., collective strolling (*jiti sanbu*) and collective tea-drinking (*jiti hecha*) – have also been introduced, all of which were peaceful but illegal. See Liu Guanglei, "Xinshiqi quntixing shijian de yufang yu chuzhi [Preventing and Containing Collective Protests of the New Period]," *Lilun dongtai* [*Theory Dynamics*], No. 1837 (December 30, 2009), p. 22.

23 See the Research Office of the Ministry of Public Security, "Woguo dangqian quntixing shijian de xianzhuang yu duice [The Current Situation and Measures Against Collective Protests in China]," in Zhongyang zhengfa weiyuanhui (ed.), *Weihu shehui wending diaoyan wenji* [*Collected Materials on Maintaining Social Stability*] (Beijing: Falu chubanshe, 2001), pp. 46–48. See also The Task Force Team (ed.), *Zhongguo quntixing tufa shijian*, pp. 12–13, 142.

24 For reports on protests participated by PLA soldiers and veterans, see *South China Morning Post*, August 17 and November 7, 2005; and Liu Guanglei, "Xinshiqi quntixing shijian de yufang yu chuzhi," p. 21.

25 Tong Yanqi and Lei Shaohua, "Large-Scale Mass Incidents in China," *East Asian Policy* (April–June 2010), p. 24.

26 Wu Shaofen, "Zhongguo shehui gongzheng de xianzhuang yu qushi," in The Task Force Team (ed.), *Zhongguo quntixing tufa shijian,* pp. 227–236.

27 Hubei Provincial Party Committee, "Guanyu yufang he chuzhuo quntixing shijian de diaoyan baogao [Research Report on Preventing and Countering Collective Protests]," in Zhongyang zhengfa weiyuanhui (ed.), *Weihu shehui wending diaoyan wenji*, p. 221.

28 Yan Junjie, *Gongan jiceng jichu gongzuo lunwenji* [*Collected Essays on the Public Security Work at Grassroots*] (Beijing: Zhongguo renmin gongan daxue chubanshe, 2003), p. 259.

29 Mu Muying, "Zhou Yongkang chengren guanbi minfan [Zhou Yongkang Admits that Popular Protests Were Caused by Officials]," *Zhengming-dongxiang* [*Contend-Trend Magazine*] (February 2009), p. 10.

30 Yang Hede, *Qunti shijian yanjiu* [*A Study of Collective Incidents*] (Beijing: Zhongguo renmin gongan daxue chubanshe, 2002), p. 7; and The Task Force Team (ed.), *Zhongguo quntixing tufa shijian*, pp. 67, 92.

31 Zhongyang zhengfa weiyuanhui (ed.), *Weihu shehui wending diaoyan wenji*, pp. 3, 11. In coping with a protest in Zhongmou County in Henan in 2004, even a local martial law had to be imposed. See Audra Ang, "Rioting in China Town Kills Seven People," *The Guardian*, November 1, 2004.

32 Cai Xia, "Qinfang guojia yu shehui guanxi nishi fazhan," p. 4; Xie Wuzhong, "Guofang dongyuan xitong chuzhi quntixing tufa shijian wenti xiantan," p. 25; and Mo Jihong, "Jinnianlai tufaxing shijian chuzhi zhong cunzai de wenti yuanyin ji jiejue duice [Problems, Causes and Solutions for Accidental Incidents in Recent Years]," *Lingdao canyue*, No. 26 (September 15, 2009), pp. 15–16.

33 Guojia jihua weiyuanhui Hongguan jingji yanjiuyuan, *Woguo shehui xingshi genzong fenxi yu duice yanjiu* [*In-depth Analysis of China's Social Conditions and Measures*] (Beijing: Macro-Economic Research Institute, December 2000), p. 97; and Zhongyang zhengfa weiyuanhui (ed.), *Weihu shehui wending diaoyan wenji*, p. 10.

34 Jae Ho Chung, Hongyi Lai and Ming Xia, "Mounting Challenges to Governance in China: Surveying Collective Protestors, Religious Sects and Criminal Organizations," *China Journal*, No. 56 (July 2006), p. 21. A similar finding is reported in Tong and Lei, "Large-Scale Mass Incidents in China," p. 24.

35 See Zhang Feng, "Quntixing shijian yisu bu yidu [Collective Protests Are Easier to Disperse than to Prevent]," *Gaige neican*, No. 8 (2009), pp. 28–29.

36 See "2006 zhongyang yihao wenjian sanpan [Three Looks at Central Document No. 1 of 2006]," *Lingdao juece xinxi* [*Information for Leaders' Decision-Making*]," No. 2 (January 2006), pp. 4–5; and The Task Force Team, "Jingji jiegou tiaozhengqi quntixing shijian fazhan qushi tezheng ye yufang," pp. 2–3.

37 Zhang Wancheng, *Nongcun zhian yu goujian hexie shehui* [*Public Security and the Construction of Harmonious Society in the Countryside*] (Beijing: Zhongguo renmin gongan daxue chubanshe, 2008), pp. 259–260.

38 The Task Force Team (ed.), *Zhongguo quntixing tufa shijian – chengyin ji duice*, pp. 53, 85–87, 91, 93. See also Kelly Chan, "Beijing Advises Local Officials to Lend an Ear to Petitioners," *South China Morning Post*, May 2, 2009.

39 The Task Force Team (ed.), *Zhongguo quntixing tufa shijian – chengyin ji duice*, pp. 272–276, 299–313; and "China Issues Statute on 'Large-Scale Mass Activities,'" *Xinhua*, September 21, 2007 at http://news.xinhuanet.com/english/2007–09/21/content_766524.htm; Jiang Xun, "Zhonggong miandui guanzhi weiji [China Faces a Governance Crisis]," *Yazhou zhoukan*, December 23, 2007, p. 32; and The Writing Team, *Yingdui tufa shijian*, Appendix-1 (pp. 184–185).

40 See *South China Morning Post*, July 14, 2009.

41 The Task Force Team (ed.), *Zhongguo quntixing tufa shijian*, pp. 55, 77, 81, 96, 107; and "Zhiyao yinfa quntixing shijian jiu mianzhi [Causing a Collective Protest Is a Cause of Removal]," *Lingdao juece xinxi*, No. 2 (November 2006), p. 14. As a matter of fact, a similar responsibility system proved very effective in China's battle against Falungong during 1999–2001.

42 Some even dub such a tendency as an "illusion of instability" (*buwending huanxiang*). The Task Force Team of Qinghua University, "Weiwen zhiduhua zhuangui zhong de fengxian yingdui [Coping with the Risk of Carrying out the

Task of Institutionalizing the Maintenance of Stability]," *Gaige neican*, No. 15 (2010), p. 13.

43 Li Qianwu, *Zhengzhi fazhan yu shehui wending* [*Political Development and Social Stability*] (Shanghai: Xuelin chubanshe, 2006), pp. 173–174; and Liu Zhenhua, *Nongcun diqu shehui zhian fangkong tixi yanjiu* [*Study of the Public Security and Control System in the Rural Society*] (Beijing: Zhongguo renmin gongan daxue, 2008), p. 44.

44 See Zhan Wancheng, *Nongcun zhian yu goujian hexie shehui*, p. 254; and Guan Wujun, "Tianjia weiwen bushi changjiu zhi ji [Maintaining Stability at Whatever Cost Is Not a Reasonable Option]," *Gaige neican*, No. 15 (2010), p. 7.

45 According to a source, while the head of this group is Zhou Yongkang, Xi Jinping is allegedly also involved in this. The head of the administrative office is Liu Jing, Deputy Minister of Public Security. Interview in Beijing in November 2009.

46 Interviews in Beijing in December 2009.

47 See Liu Zhenhua, *Nongcun diqu shehui zhian fangkong tixi yanjiu*, p. 49. As of 2003, Hunan ranked fourth in the frequency and participant number of collective protests. See Chung, Lai, and Xia, "Mounting Challenges to Governance in China," p. 21.

48 Zhan Wancheng, *Nongcun zhian yu goujian hexie shehui*, pp. 23–25; and Liu Zhenhua, *Nongcun diqu shehui zhian fangkong tixi yanjiu*, pp. 46, 49.

49 Liu Zhenhua, *Nongcun diqu shehui zhian fangkong tixi yanjiu*, p. 51.

50 The stability preservation offices are replicated along the administrative hierarchy down to the township level. Their main function is to plant "eyes and ears" on local activities, which constitute the vertical surveillance network against collective protests and other similar threats. Interviews in Beijing in November 2009.

51 See *Renmin ribao* [People's Daily], October 15, 1999; *New York Times*, December 23, 1999; *South China Morning Post*, June 8, 2000; *Business Week*, August 8, 2000; and *Washington Post*, December 28, 2000. See Clark Boyd, "Bypassing China's Net Firewall," at http://news.bbc.co.uk/go/pr/fr/-/2/hi/technology/3548035.stm (last accessed on April 3, 2004).

52 *Renmin ribao*, November 7 and December 29, 2000.

53 See Zhongguo shehuikexueyuan, *2010nian shehui lanpishu* (Beijing: Shehuikexue chubanshe, 2009); *Nanfang dushibao*, December 22, 2009; *New York Times*, December 18, 2009. See also Qun Yan, "Yingdui wangshang qunti shijian quanyuan youdian bu shiying [Officials Are Not Used to Dealing with Online Collective Protests]," *Gaige neican*, No. 17 (2009), pp. 29–31.

54 Jonathan Ansfield, "China Starts New Bureaucracy to Curb Web," *New York Times*, April 16, 2010.

55 Liu Zhenhua, *Nongcun diqu shehui zhian fangkong tixi yanjiu*, pp. 17, 104, 199; The Writing Team (ed.), *Yingdui tufa shijian*, pp. 11, 29; and interviews in Beijing in December 2009. A similar categorization is employed in the area of managing endemic/pandemic crises. See the contribution by Hongyi Lai to this volume (Chapter 5).

56 Yu Jianrong, "Dangqian bufen diqu nongcun quntixing shijian de zhuyao tedian [Major Characteristics of Collective Protests in Some Rural Areas]," *Lingdao canyue*, No. 21 (July 25, 2002), pp. 14–15; and Zhan Wancheng, *Nongcun zhian yu goujian hexie shehui*, p. 255. As a matter of fact, since 2001, all major cities in China were ordered by Beijing to set up anti-riot police to deal with such contingencies. Interviews in Beijing in November 2009.

57 See Dennis J. Blasko, *The Chinese Army Today: Tradition and Transformation for the 21st Century* (London: Routledge, 2006), pp. 17, 66; and Xu Shihong, "Shenhua wujing jiazhi renshi [Deepen Our Understanding of the Value of PAP]," *Guofang*, No. 2 (2008), pp. 66–67.

58 Chinese sources take pains to characterize most of the collective protests as "contradictions among the people" as opposed to those between the people and the enemies

of the people. The deployment of the PLA is generally designed to deal with the latter. See, for instance, Task Force Team (ed.), *Zhongguo quntixing tufa shijian – chengyin ji duice*, pp. 2, 5, 127.

59 See David Shambaugh, *Modernizing China's Military: Progress, Problems and Prospects* (Berkeley: University of California Press, 2002), p. 22.

60 Blasko, *The Chinese Army Today*, p. 67.

61 See The Academy of Military Sciences, *Zhongguo renmin jiefangjun gaige fazhan 30nian [Reform and Development of PLA in the Last 30 Years]* (Beijing: Jiefangjun chubanshe, 2008), pp. 371–373. See also Yang Yi (ed.), *Zhongguo guojia anquan zhanlue gouxiang [Ideas and Plans for China's National Security Strategy]* (Beijing: Shishi chubanshe, 2009), pp. 397–407. For the clause that provides a legal basis for the PLA's involvement in containing "accidental incidents," see Article 14 of the "Law on Coping with Accidental Incidents" in the Writing Team (ed.), *Yingdui tufa shijian*, Appendix-1, p. 186.

62 "Dui wanshan junzheng junmin lianhe yingji zhihui tixi de tantao [Discussing the Military-Government Joint Command System in Emergency Management]," *Guofang*, No. 3 (2008), pp. 35–36.

63 *Guofang*, No. 9 (2009), pp. 13–14.

64 Sha Zhenghua, "Jiaqiang xianji renwubu jianshe de ruogan sikao [Some Thoughts on Building up the PAP at the County Level]," and Wen Muhai and Li Zhilun, "Zhaoyan suixing duoyanghua junshi renwu xuyao dali jiaqiang minbing yubeiyi budui jianshi [Carrying out Diversified Duties Requires the Strengthening of Reserve Militia Forces]," *Guofang*, No. 12 (2009), pp. 28–29, 38–40.

65 See Lu, *Sulian xingwang shilun*; Special Editing Team (ed.), *Xingshuai zhi lu – waiguo butong leixing zhengdang jianshe de jingyan yu jiaoxun [The Road to Prospering and Perishing – Experiences and Lessons from Different Types of Political Parties Abroad]* (Beijing: Zhonggong zhongyang dangxiao chubanshe, 2002), pp. 127–178; and Chen Daichang, *Jiaqiang dang de zhizheng nengli jianshe de diaocha yu yanjiu [Research and Study on Enhancing the Party's Governance Capacity]* (Beijing: Zhonggong zhongyang dangxiao chubanshe, 2004).

66 This term was first used by Thomas P. Bernstein in his monograph *Unrest in Rural China: A 2003 Assessment* (Irvine, CA: Center for the Study of Democracy, 2004), p. 2. As for the high likelihood of stable instability as opposed to other possibilities (i.e., governed stability, democratizing instability, and disintegrative instability), see Chung, "Assessing the Odds Against the Mandate of Heaven," pp. 117–120.

67 For views that assess such prospects to be unlikely, see Harvey Nelson, "The Future of the Chinese State," in David Shambaugh (ed.), *The Modern Chinese State* (Cambridge: Cambridge University Press, 2000), pp. 230–236.

68 See Kevin O'Brien and Lianjiang Li, "Villagers and Popular Resistance in Contemporary China," *Modern China*, Vol. 22, No. 1 (1996), pp. 28–61; and Thomas Bernstein and Xioabo Lu, *Taxation without Representation in Contemporary Rural China* (Cambridge: Cambridge University Press, 2005), pp. 139–140. See also Cai Yongshun, "Managed Participation in China," *Political Science Quarterly*, Vol. 119, No. 3 (2004), pp. 425–452.

69 See John Arquilla and David Ronfeldt, *Networks and Netwars* (Santa Monica, CA: RAND, 2001), Chs 4 and 5.

3 Chinese military crisis behavior

From confrontation and conflict prevention to win–win management

Tuosheng Zhang

In the past 60 years and more since the founding of the People's Republic, with changes in the international situation and China's domestic and foreign policies, the country's military crisis behavior has also undergone profound changes. In the past two decades, the military crisis management mechanism in China has strengthened and improved with increasingly rich and developed policies and principles guiding crisis management and markedly improved capabilities in managing military crises.

This chapter attempts to analyze, induct and outline China's military crisis behavior (crisis management in particular) on the basis of major military crises China went through, sum-up the evolution of such behavior and the underlying reasons, and on that basis look into future military crisis behavior of China. The chapter consists of the following four sections: relevant definitions and objects of study, overview of Chinese handling of military crises, analysis and summary of China's military crisis behavior and future prospects.

Relevant definitions and objectives of the study

Definition of military crisis

For the purpose of this chapter, a military crisis refers to a dangerous state of possible outbreak of military conflict or even war between political groups within a state or between states and a special form of struggle between peace and war.[1] It is mainly composed of three elements: core or major interests of the relevant parties are threatened; there are time pressures or sense of urgency; there is serious risk of an outbreak of military conflict.[2] A military crisis may occur either within a country or between countries; and those between countries are international military crises.[3]

There are both potential and actual military crises. A crisis is potential when conflict of interest in military security exists between the relevant two or more parties and yet the levels of antagonism and external conditions are yet to reach a certain critical point. A potential becomes an actual crisis when antagonism increases and certain external change leads to rapid intensification and surfacing of the conflicts of interest.

Another type is called a near crisis, which usually does not lead to military conflict or war despite the existence of relatively big conflicts of military and security interest and time pressures, which may bring about serious damage to the political and diplomatic relationships of the states concerned.[4] Such a crisis is usually an accidental or unexpected event, and the possibility of accidental discharge does exist in reality. Furthermore, given the time constraint, if a misjudgement takes place, the risk of a military conflict will rise sharply.

It must be pointed out that the concept of a military crisis is not the same as that of a security crisis, with the latter having wide connotations covering both traditional and non-traditional security crises and the former basically overlapping the concept of traditional security crises. Although a military crisis may involve some non-traditional security situations such as crises caused by international terrorism or proliferation of weapons of mass destruction, it generally speaking differs from the majority of non-traditional security crises such as financial crises and energy crises and those caused by climate change, environmental pollution and communicable diseases, which might also, in extreme conditions, evolve into military crises.

Definitions of military confrontation and crisis management

There are two types of response by crisis actors to a military crisis. The response by a crisis actor determined to resolve a conflict of interest by military means so as to safeguard its own interests to the maximum degree possible is defined in this chapter as military confrontation. A crisis actor may take the initiative or be forced to take military actions in response. The military confrontation may also be the result of failure in crisis management.

A crisis actor may also work to control the risk of crisis escalation, prevent the outbreak of military conflict and gradually relax the crisis while making efforts to safeguard its own interest. This is military crisis management. In most cases, the success in managing a military crisis is reflected not in finding a solution to the conflict of interest but rather in avoiding the worst scenario, war. In the process of crisis management, the states concerned usually take various negotiation strategies and compete and cooperate with each other at the same time.[5]

Besides the above-mentioned two differing crisis behaviors, there sometimes appears a third situation in which managing a crisis is the dominant approach, supported by military confrontation. In this situation, the extent of military confrontation is usually rather limited and aimed at realizing management of the crisis.

Objectives of the study

Per the above definitions, first, this chapter discusses the major international military crises (including actual, potential and near crises) experienced by the PRC. These crises took place either directly between China and another party or between other countries but with China involved due to its major bearing on China's national security interest. Except for crises across the Taiwan Strait, domestic military crises in China are not discussed.[6]

Second, since 1949, some military crises that China experienced directly caused military conflict or war between China and other countries; some witnessed both military confrontation and crisis management in their development processes; and some were managed, brought under control and mitigated. This chapter touches upon the two major military crisis behaviors of China: military confrontation and crisis management, with a focus on the latter. The confrontational part is discussed for the purpose of a better illustration of the major change in the military crisis behavior of China.

Third, the various non-traditional security crises that differ significantly from a military crisis are not covered in this chapter, which, however, does look into crises of nuclear proliferation in the world, close to China, in particular. Since the beginning of the new century, the crises and challenges brought about by international terrorism have gained increasing attention from China. Up to now, China has rarely involved itself in managing international terrorist events. In the long term, however, gradual intervention will hardly be avoidable for China.

Fourth, besides the concept of military crisis management, this chapter uses also the concept of security crisis management since decision-making and management of military crises in China have always been a component of its security decision-making and management of security crises. In some cases, the latter concept is more accurate and reflects better the reality in China.

Overview of Chinese handling of military crises

During the Cold War

The Cold War (1946–1991) had already started when New China was founded in 1949. In the 40-plus years, China encountered many military crises and adopted or was forced to adopt military means to resolve them. The Indochina War was the first war in which the PRC intervened. Shortly after the end of World War II, the Indochinese Communist Party led by Ho Chi Minh launched a war against French colonialists and requested China in early 1950 to send troops to Vietnam. Proceeding from the position of supporting national liberation movements and eliminating threats of Western colonialist forces in the surrounding regions, Chinese leaders did not hesitate to decide on aiding the Vietnamese in their resistance against France. However, after careful consideration, they decided not to send troops to be directly involved in the war but rather to send a military advisory team and provide weapons and materials assistance. In 1953, the Indochina War became a game of two halves. France attempted to seek compromise whereas the United States hoped that France would continue fighting and actively planned to send troops in to help the French. In this situation, China decided to take an active part in the Geneva Conference and worked on the parties to the conflict, the Vietnamese in particular, making an important contribution to the signing of agreement on the cessation of hostilities in July 1954.[7]

The 1950–1953 Korean War was a major regional war with direct involvement of China. The American military intervention turned the Korean civil war into an international war. Chinese leaders decided to make preparations on two fronts, on the one hand ordering the assembly of troops towards the Sino-Korean border, and on the other engaging in a series of diplomatic efforts to reduce tension and avoid expansion of war or direct involvement in the war.[8] Since communication between the two sides was very difficult due to absence of diplomatic ties, and the two sides completely lacked understanding of each other,[9] a major military contest, which had brewed for a long time but both sides had hoped to avoid, finally became real when massed US troops marched across the thirty-eighth parallel. This war was of great and far-reaching influence on strengthening military crisis management between the two countries later.

In 1954 and 1958, China waged two military struggles against Taiwan in the Taiwan Strait known as "bombardment of Quemoy." The objectives of the first one (from August 1954 to May 1955) were to liberate coastal islands of Zhejiang Province, counter harassing attacks by the Chiang Kai-shek army, and prevent the United States and Taiwan from signing the military alliance treaty. The second campaign (from August to October 1958) was designed to strike back at harassment by Chiang troops, capture coastal islands of Fujian Province occupied by Chiang when opportunities permitted, and get a clear picture of the scope of the US–Taiwan Joint Defence Treaty.[10] During both crises, the United States responded strongly by executing a massive assembly of troops to the Taiwan Strait and announcing a nuclear threat to China. In the tense situation, Chinese leaders made effort to avoid direct military conflict with the United States[11] while conducting military operations against Taiwan. The military action was ended when some of the objectives had been reached and reflection had been given to policies guiding struggles against Taiwan and the United States.[12]

In 1962, a severe military conflict broke out between China and India due to territorial disputes.[13] Armed personnel of the two countries had the first exchange of fire at the border in autumn 1959. To relax tension, China unilaterally withdrew 20 kilometers from the line of actual control and in the following year put forward specific proposals to India concerning the signing of a border agreement. India did not make a positive response to China's self-constraint. Since the end of 1961, India started to nibble at Chinese territory by dispatching soldiers and setting-up posts in the disputed western section of the border, and in mid 1962, the Indian army went further to cross the line of actual control in the disputed eastern section to build-up military posts.[14] China proposed to negotiate again, to which India responded that only the "withdrawal of Chinese army" should be discussed. In this situation, Chinese leaders judged that India intended to make use of the serious difficulties facing China[15] to forcefully change the status quo of the border by military encroachment. To restore the status quo of the border and maintain peace in border areas, China decided to carry out counter-attack. On 20 October, the People's Liberation Army attacked invading Indian armies. In late November, China declared a ceasefire and subsequently withdrew its army to 20 kilometers within the Chinese side of the line of actual control.

Aiding Vietnam to resist US aggression brought China to a military conflict with the United States, which was to be the longest during the Cold War. The United States comprehensively escalated its warfare in Vietnam as of August 1964 by fighting a ground war in the South and bombarding the North. China made a strong response by announcing the readiness to dispatch personnel to Vietnam to fight American troops and to strike US military aircraft invading Chinese territorial air space.[16] Meanwhile, China still hoped to avoid direct large-scale military conflict with the United States and prevent the United States from bringing war into China. In the first half of 1965, Chinese leaders conveyed messages to the United States through various channels: China had no intention to wage a war against the United States; Chinese reactions would be reciprocal, i.e., it would not send troops to South Vietnam so long as the United States did not cross the seventeenth parallel; and Chinese troops would only offer combat assistance in the North if the American military operations in the North were limited to naval and air operations.[17]

This time, the United States apparently noticed the Chinese signals. Until the end of the war, the United States never launched a ground war in North Vietnam and always strictly prohibited its military aircraft from entering into Chinese territory. A direct military confrontation between China and the United States was avoided. Military conflicts between China and the United States gradually stopped after 1968 as the war gradually "Vietnamized." The battle over Xisha Islands in January 1974 was an interlude in China's assistance of Vietnam to resist United States aggression. In the then-favorable situation, Chinese leaders decided to counter-attack the South Vietnamese navy, which had invaded Xisha Islands, and recovered islands previously occupied by South Vietnam.

Border military conflict broke out between China and the USSR in 1969. The direct trigger was a boundary dispute and the deeper cause was intensification of ideological differences and relationship breakdown between the two countries. Before the outbreak of conflict, serious friction had occurred in the eastern section of the China–Soviet boundary and military conflict could be triggered at any moment. Given the Soviet pressure, Chinese leaders decided to take a tit-for-tat and no-giving-in policy. Border forces of the two sides had skirmishes on Zhenbao Island in March and in Tielieketi, Xinjiang, in August 1969.[18] In face of the severe situation, Mao Zedong called on the whole country to "prepare for war" and China entered into combat readiness. The Soviet Union speeded-up projection of troops to the border areas, doubling the number of troops there in a short period, and threatened to use nuclear weapons against China. Although a meeting between the two premiers at Beijing Airport in September to a certain extent relaxed the confrontational atmosphere, the agreement reached during the meeting to relax tension and maintain border status quo was not supported by top leaders of both countries, leaving the meeting results invalid. After that, Sino-Soviet relations were for a long time in full confrontation.

The last military conflict China engaged in during the Cold War took place in 1979. As early as around the time of Vietnamese unification in 1975, historical grievances, territorial and maritime disputes and divergence of strategic interests between China and Vietnam began to surface. The Vietnamese invasion of Cambodia in 1977 led to drastic deterioration of China–Vietnam relations. In 1978, with Soviet support, Vietnam launched a massive invasion of Cambodia, igniting the fuse of the China–Vietnam military crisis. In order to maintain its own security and regional security and oppose Soviet and Vietnamese expansion in Indochina, Deng Xiaoping and other Chinese leaders in autumn 1978 decided to carry out a self-defence war against Vietnam. To support the military campaign, China conducted active diplomatic moves[19] and made full preparation for possible Soviet intervention. The war broke out in February 1979 and the Chinese troops withdrew from Vietnam the following month after reaching basic operation targets in Vietnam. Starting from 1980, small-scale cross-boundary combats between the two countries continued for quite some years.[20] The two countries also fought in Nansha Islands (Spratly) in March 1988, in which China recovered a few islands and reefs previously occupied by Vietnam.

Besides, China also intervened during the Cold War to a certain extent in the two Indo-Pakistani Wars in 1965 and 1971. On both occasions, China provided military assistance to Pakistan upon the latter's request. In 1965, Chinese leaders even made the decision to send troops in assistance of Pakistan if necessary. Chinese troops were not sent since India backed-off on the Sino-Sikkim border after receiving China's "note of ultimatum."[21]

Since the end of the Cold War

Since the end of the Cold War, military crises have taken place from time to time in the region and the world at large. However, direct military crises between China and other countries have markedly decreased. China has not been involved in any external military conflict and has gradually become an important force in multilateral international crisis management.

The Meiji Jiao (Mischief Reef for the Philippines) incident in 1995 epitomized China's disputes with five Southeast Asian countries in the South China Sea. In January that year, a Chinese scientific research ship was sent to Meiji Jiao to conduct scientific research and build wind shelters for the fishermen. The Philippine government moved to increase troops to the disputed area on grounds that a "Chinese military ship" had "invaded" its sea area and built a "military facility" on its reefs and made a series of moves.[22] Other Southeast Asian countries having a stake in disputes in the South Sea got involved by expressing serious concern over the Meiji Jiao incident and intensifying their own activities in the Nansha area. The incident was for a time the focus of international attention. On the one hand, the Chinese government made diplomatic representations to the Philippine government, opposing its attempt to internationalize the Nansha dispute. On the other, China increased alerts in the relevant sea areas in a firm bid to safeguard China's legitimate rights and interests. At the China–ASEAN foreign ministers' dialogue at the end of July, China put forward comprehensive propositions to resolve the Nansha disputes and expressed its willingness to peacefully resolve disputes with relevant countries according to the principles of international law and the United Nations Convention on the Law of the Seas, which was welcomed by parties to the conference.[23] Then tension caused by the Meiji Jiao incident gradually relaxed.

In 1995 and 1996, the third Taiwan Strait crisis broke out across the strait and between China and the United States. The crisis was caused by the US Administration approving, in violation of its commitment, a visit by Taiwanese leader Lee Teng-hui to the United States.[24] China reacted strongly by first interrupting all dialogues and exchanges with the United States, followed by recalling its ambassador to the United States and then conducting missile test launches and military exercises in the Taiwan Strait in June and July. In March 1996, in light of the so-called "presidential" election in Taiwan, China declared a further missile test launch and military exercise. The United States had an urgent discussion about military deployment in the Taiwan Strait (and later dispatched two aircraft carrier formations).

In the face of the severe situation, China completed the test launch and conducted a larger-scale joint military exercise of the army, navy and air force on 28 March, the day on which the presidential election in Taiwan was held and the United States aircraft carrier Nimitz arrived in the West Pacific. The crisis contest gave the United States new understanding and knowledge about the sensitivity of the Taiwan question,[25] containing to a certain extent the momentum of Taiwan's attempt to change the one-China framework by "pragmatic diplomacy." During this crisis, since both China and the United States wished to avoid military conflict and the two sides maintained necessary communication,[26] no real military face-off between China and the United States took place.[27]

After that, pro-independence forces created trouble many times, causing tension across the Taiwan Strait.[28] In response, China formulated the Anti-Secession Law to explicitly declare the bottom line for the use of force against Taiwan independence while fully demonstrating good will towards people in Taiwan. At the same time, it strengthened preparation for necessary military struggles. On the basis of maintaining status quo in the strait, China made active effort to develop anti-Taiwan-independence cooperation with the United States and main opposition parties in Taiwan. The Chinese endeavor formed strong pressure over pro-independence forces in Taiwan, whose plots to instigate a crisis across the Taiwan Strait were again and again frustrated.

At the turn of the century, two unexpected military crises occurred between China and the United States due to the incidents of embassy bombing[29] and aircraft collision.[30] Both were actually near-crises. In comparison, the embassy bombing incident was more unexpected and more serious[31] than the aircraft collision, while the latter was more complicated[32] and contained more risks[33] than the former. Although lacking experience in handling such contingencies,[34] Chinese leaders fairly rapidly adopted measures in response and gradually relaxed the two crises. The basic policies of China were to firmly safeguard national sovereignty and dignity against wrongdoings of the United States and at same time make effort to maintain the overall interest of China–US relations and avoid confrontation. The embassy bombing incident moved to relaxation after multiple apologies and compensation by the United States. The aircraft collision incident was finally resolved by separating the handling of the crew and the plane.[35] The two incidents gave Chinese leaders valuable experience in handling emergencies. Since then, crisis management and coordination mechanism in China has markedly improved.

Due to disputes over maritime rights and interests, a crisis also broke out between China and Japan in 2004. In May, the Japanese media began to cover extensively China's exploration of oil and natural gas in the East China Sea, claiming that China's exploration effort to the west of the median line would suck away Japanese resources, accusing China of attempting to take all submarine resources in the East China Sea and demanding that the Japanese government take drastic measures to safeguard Japan's interests.[36] The Japanese government then increased its military surveillance and resource investigation in the relevant sea area and made "solemn representations" with China. Disputes between the two countries over oil and gas development and demarcation of boundary in the East China Sea suddenly heated up. The East China Sea dispute added tension to the already rather confrontational public opinions and tense bilateral relations.[37] The idea that there would definitely be a war between China and Japan appeared in the media of both countries. In the extremely tense situation, the Chinese government insisted on its sovereign rights and firmly refused Japan's unreasonable demands. Meanwhile, it also proposed to Japan to conduct joint development of the disputed areas and to have a dialogue over the dispute. Furthermore, interagency coordination and guidance of public opinions were strengthened so as to prevent inadvertent events. In October, the two countries started their consultation on the East China Sea and tension was brought under control.

Since 1998, China has also intervened in international nuclear crises, making active efforts to safeguard the international nuclear non-proliferation regime, relax tension and avoid military conflict. During the Indian–Pakistani nuclear crisis,[38] China cooperated fully with other members of the UN Security Council,[39] bringing the crisis under control fairly rapidly. In the face of the DPRK (Democratic People's Republic of Korea) nuclear crisis,[40] China has maintained close contact and consultation with concerned parties including the DPRK and the United States, taken the Six-Party Talk as the main platform to resolve the DPRK nuclear issue and at the same time supported the UN Security Council in playing a necessary role.[41] The Chinese government also strengthened its internal coordination among different departments and military preparedness. As to the Iranian nuclear crisis,[42] from discussions at the International Atomic Energy Agency Board of Governors' meetings to consultations at the UN Security Council to the six-nation dialogue and to directly working on Iran, China's involvement in managing the crisis has deepened and its influence increased in the past few years.[43]

Besides the above-mentioned nuclear crisis, China also made unremitting efforts to mitigate the Iraqi crisis[44] before the outbreak of the Iraq War in 2002. At that time, China was an active participant in international counter-terrorism cooperation but was not in favor of the United States waging an anti-terror war in the world, and opposed US "pre-emptive" military strikes against other countries on the grounds of counter-terrorism. In the high tensions when the United States and United Kingdom were determined to use force against Iraq, China called for a political solution within the framework of the United Nations and opposed the UN Security Council automatically authorizing the United States to use force against Iraq.[45] To this end, China worked with many other countries in wrestling with the United States and United Kingdom. Even though the efforts in the end did not prevent the United States and United Kingdom from launching the war, they upheld international justice and safeguarded the basic principle of the UN Security Council, maintaining peace and security, and in the end the Security Council did not give a green light to the use of force against Iraq by the United States and United Kingdom. During the crisis, China coordinated and collaborated with France, Germany, Russia and other countries holding similar views and engaged in a diplomatic struggle with the United States, United Kingdom and other countries of opposite opinions. It relied on the Security Council as the main platform of diplomacy and actively employed other means such as bilateral consultation and leaders' diplomacy. These practices accumulated new experiences for China to conduct international crisis management.

Analysis and summary of Chinese military crisis behavior

Changes in Chinese military crisis behavior

The above overview of China dealing with military crises suggests that in the past 60 years the trajectory of Chinese crisis behavior has been one from military confrontation to crisis management and from conflict avoidance to striving for win–win.

During the Cold War, China directly adopted military means on five occasions to deal with military crises, including aiding Vietnam against France, Sino-Soviet conflict on Zhenbao Island, fighting with Vietnam (South Vietnam) over Xisha, self-defensive war against Vietnam, and fighting with Vietnam (North Vietnam) over Nansha. In all these cases, military confrontation was China's expressed choice.

In the same period, there were also five cases in which China used both military confrontation and crisis management, including assisting DPRK to resist the United States, two military struggles in the Taiwan Strait, aiding Vietnam to resist the United States, and the Sino-Indian border conflict. Before getting involved in the Korean War and engaging in a military conflict with India, China actually made efforts to mitigate the crises without success. However, in striking against Taiwanese troops and limited intervention in the Vietnam War, China was able to avoid direct military conflict with the United States, achieving certain success in crisis management. In these cases, military confrontation remained an important option for China.

With the above-mentioned two categories combined, China intervened or was involved in ten military conflicts and wars during the Cold War. Besides, China was close to intervening in the first Indo-Pakistani War in 1965. Military confrontation supported by some crisis management effort was the main feature of Chinese military crisis behavior during that period.

Various military and security crises continue occurring since the end of the Cold War. However, China has not engaged in any external military conflict. The various crises directly involving China, such as the Meiji Jiao incident with the Philippines, the third Taiwan Strait crisis, the embassy bombing, the aircraft collision and East Sea crisis with Japan, were all well controlled. On the Taiwan question in particular, although China reserves its right to use force, it has extended crisis management from between China and the United States to across the strait, demonstrating the greatest determination and desire for peaceful reunification.

Meanwhile, major disputes over land borders or maritime rights and interests that had in history led to armed conflict have been either fundamentally resolved, for example, the Sino-Soviet/Russian border dispute,[46] Sino-Vietnamese disputes on land borders and border demarcation in the Beibu Gulf,[47] or brought well under control, for example, the Sino-Indian territorial dispute[48] and Sino-Vietnamese dispute in the South China Sea, without causing any military crisis.

China has also intervened in the management of major international security crises such as the Indian-Pakistani nuclear crisis, Iraq crisis, DPRK nuclear crisis and Iranian nuclear crisis by making active efforts to control crisis escalation, avoid military conflict and seek comprehensive long-term resolution to the conflict of interest among concerned parties. China has no direct involvement in these crises. Nonetheless, as these crises, mostly occurring in China's surrounding areas and in a region where China has a stake, pose major threats to regional and world peace and breed huge risks of military conflict or even war, China has increased its intervention along with increased national strength and international influence.

In short, since the 1990s, outstanding changes have taken place. Military confrontation, which once was the dominant crisis behavior during the Cold War, has been replaced by crisis management. External military conflict, a recurrent result of crisis during the Cold War, has been avoided with active crisis management. With strengthened crisis management, the old trend of potential crisis developing into actual crisis has been changing with some potential crises weakened and resolved and some others under control. Striving for a lasting solution to conflict of interest and win–win result has become the direction of Chinese efforts.

As a matter of fact, the starting point of the major changes in Chinese crisis behavior can be traced to an earlier point in time, that is, early 1980s. In the decade since then to the end of the Cold War in 1991, the number of military crises involving China greatly decreased, except for small-scale conflicts with Vietnam. Before the outbreak of the Gulf War in 1990, China had conducted active diplomatic mediation in pursuit of a peaceful settlement to the Gulf crisis by Iraq withdrawing from Kuwait,[49] which marked an important step towards involvement in international crisis management.

Development of and changes in Chinese military crisis management mechanism

The military crisis management mechanism of China has always been determined by its leadership structure and operations in national security. Since the founding of New China, the national security leadership structure has in general been relatively stable and consistent. The Political Bureau of the CPC Central Committee and its Standing Committee are responsible for major diplomatic and military decisions. The Secretariat of the CPC Central Committee handles the daily work of the Central Committee, including matters related to national security. The State Council, Central Military Commission, relevant central leading small groups, Ministries of Foreign Affairs and National Security participate in the handling of specific affairs related to national security.[50]

The operations of the above national security leadership structure, however, experienced three big changes in the past decades, with their respective impacts on the nation's military crisis management mechanism. The first was during the 10-year Cultural Revolution when the party and government organizations at all levels were paralyzed and performed practically no function and the original collective leadership was seriously undermined. As a result, decision-making was controlled by a handful of people and ultimately up to Mao alone. In that situation, military crisis decision-making also suffered big adverse impact. Then with the end of the Cultural Revolution, normal operation of national security leadership structure was restored and collective leadership with Deng Xiaoping at the centre was formed. Accordingly, the military crisis management mechanism was again strengthened. The last major change occurred along with veteran leaders stepping down from the central political stage. Decisions are no longer made on the basis of experiences of a few leaders. Rather, they are made by the collective top leadership and increasingly institutionalized.[51] Such a transformation has speeded-up the strengthening and improvement process of the Chinese military crisis management mechanism.

Since the late 1990s, the Standing Committee of the Political Bureau of the CPC Central Committee has held the ultimate decision-making power and the top Party and state leader plays a central role in major security and diplomatic decision-making. In the face of various military and security crises, the Central Foreign Affairs Leading Small Group, Central National Security Leading Small Group,[52] Central Taiwan Affairs Leading Small Group and Central Military Commission are responsible for formulating relevant policies.[53] The Ministry of Foreign Affairs, General Staff Department, Central Foreign Affairs Office and State Council Taiwan Affairs Office are responsible for internal coordination, formulation of emergency response plans and putting forward policy recommendations. Various other relevant departments and think-tanks are also involved in policy coordination.

Meanwhile, various means of external communications such as hotlines, emergency liaisons procedures and dispatch of special envoys have gradually been put in place and improved (although it will still take time for their effective use in crises to be recognized and implemented). For example, China and the United States have set up a hotline between heads of state and military hotlines, China and Russia have a hotline between heads of state, China and Japan have one between the prime ministers and are discussing the establishment of naval and air force emergency liaison procedures, and China and the Republic of Korea now have direct telephone between their navies and air forces. Special envoys were sent to handle crisis across the Taiwan Strait, DPRK nuclear crisis and some other international crises. Given the increased application of information technology and rising influence of public opinions, the Chinese government is paying more and more attention to listening to and guiding public opinions. The relevant mechanisms that have been put in place have played important roles in managing the aircraft collision incident with the United States, cross-Taiwan-Strait tensions, political and security frictions with Japan, the DPRK nuclear crisis and some other security differences with other countries.

Furthermore, in the past two or three decades, remarkable progress has been made in the information and intelligence field, an important component of national security decision-making and military crisis management mechanisms. The capacity of the diplomatic, media, military and security departments to collect information and intelligence has grown markedly.[54] The relevant leading organizations are also more capable of analyzing and evaluating information and intelligence. This has enabled Chinese leaders to make more timely and accurate decisions concerning various military and security crises.

Development of and changes in the principles and behavioral paradigms of Chinese crisis management

In the past 60 years, Chinese leaders have, in the rich practices of dealing with military and security crises, formed some basic principles and behavioral paradigms of crisis management, which are illustrative of Chinese military crisis management.

The crisis management principles first came from some important strategic and tactical thoughts formed by Mao Zedong and other veteran leaders in the long revolution. For example, the principle of "strategically taking the enemy lightly and tactically taking the enemy seriously" requires a combination of strategically adhering to principles with tactically showing flexibility and of daring to fight with being good at fighting. The principle of being "on just ground, to one's advantage and with restraint" stresses the defensive, local and temporary nature of a struggle on the basis of owning moral high ground so as to unify self-defence, victory and truce. The principle of being "both-handed" or "talk for talk and fight for fight" suggests the necessity to combine soft means with tough ones and to struggle while seeking cooperation. That "promise must be kept and action must be resolute" suggests the necessity to be equal of one's words once the determination and bottom line are declared. These strategic and tactical thoughts have been extensively used in dealing with various military and security crises since the founding of New China, continue play an important role of guidance in crisis management today and are recognized by the Chinese strategic research community as crisis management principles with special Chinese features.

With the changing international security situation and enhanced understanding of the dangers of military and security crises and the significance of crisis management, Chinese crisis management principles are also developing and changing, with their contents enriched and improved. Since the end of the Cold War, in the process of handling external relations, including Chinese–foreign military crises and other international crises, Chinese leaders have put forward some new concepts and principles guiding the handling of military and security crises as the following: sticking to the United Nations Charter and norms of international law; except for self-defence, the use of force against another country must be authorized by the UN Security Council; mastering the big picture of overall state-to-state relations; seeking peaceful dialogue, increase of trust and avoidance of confrontation as first priorities; and dual focus on crisis prevention and crisis control.[55] Chinese academia has also conducted research into the basic principles of crisis management, the results[56] of which have won recognition from competent government departments and played a role in supplementing and giving details to the guiding principles of crisis management in China.

The above crisis management principles are not all at the same level and the author has no intention to categorize them. But one thing is for sure: China has registered remarkable progress in developing military and security crisis management principles, with both continuity and innovation on the basis of Chinese and foreign experiences. The starting point of crisis management has shifted from relying on struggle to giving consideration to both struggle and compromise.[57]

Compared with the development and changes of military and security crisis management principles, the relevant behavioral paradigm[58] of China has been relatively stable in the past 60 years. The overview of Chinese handling of military crises reveals that some basic actions appeared repeatedly, including sending out diplomatic warnings in a timely manner,[59] demonstrating credible deterrence by adopting certain military actions,[60] striking only after the enemy has struck and not firing the first shot, seeking necessary compromise in order to safeguard long-term and overall interest and always valuing morality and "face,"[61] and so on.[62]

Within the last decade, with the evolution of military and security crisis management mechanisms and guiding principles, the relatively stable military crisis management paradigm has also gone through some delicate yet important changes, the most significant of which are as follows: first, warning signals are increasingly sent by non-military means such as diplomacy rather than military actions. Second, more attention is given to observing the relevant international laws and seeking legitimacy of action. Third, efforts have been redoubled to establish military and security confidence-building measures with the concerned parties. Fourth, China is increasingly ready to take the initiative to seek mutual compromises and concessions, striving for win–win and avoiding lose–lose. These new features have made the country's military and security crisis management more rational and effective and rendered its crisis behavior more predicable.

Reasons behind changes in China's crisis behavior

The crisis behavior of China is not a random choice of leaders but rather determined by many international and domestic factors, the most influential of which being international strategic patterns, relations with the international systems, the central tasks and ideology of the CPC and the state and China's outlook of international security order[63] formed thereof.

Generally speaking, during the Cold War in which the United States and the Soviet Union competed for hegemony and in particular in the first 30 years since victory of the Chinese revolution, China was in serious hostility first with the United States and then with the Soviet Union, entered into alliance and was for a long time excluded from major international organizations such as the United Nations.[64] It was believed in China that the world was still in an era of revolution and war and that a new world war would be hardly avoidable. The predominant polices were to take class struggle as the focus of work and to oppose imperialists, revisionists and reactionaries. With all of these factors plus the special experience of Mao Zedong and other Chinese leaders of gaining national independence and liberation through long-term revolutionary war and their unique view of crisis,[65] it was nothing strange to see New China intervening or involved in one military conflict after another. At that time, subjectively, in the minds of Chinese leaders crisis management was not a major means to deal with crises, and efforts to avoid military conflict produced only limited results due to lack of objective conditions. Military struggle playing a predominant role and crisis management auxiliary was the main feature of Chinese crisis behavior during that period.

In the final years of the Cold War and since the end of the Cold War in particular, with development of multi-polarization, globalization and "informationization," China's international and surrounding environments have gone through major changes. Bipolar confrontation disappeared, peace and development became the theme of the era and profound changes took place in the forms and role of war.[66] With relations among major powers showing a new pattern[67] and relations with neighbors improving, China was no longer directly confronted with major military threats. At home, China gave up the wrong policy of taking class struggle as the focus of work, and reform and opening-up for modernization continued moving forward. Given these big changes, Deng Xiaoping believed that world wars can be avoided and that China should pursue an independent foreign policy of peace and a policy of non-alignment,[68] seek peaceful development and actively integrate into the international community. Since the start of the new century, Chinese leaders such as Jiang Zemin and Hu Jintao have explicitly put forward a new security concept[69] centring on mutual trust, mutual benefit, equality and cooperation, and expressed the desire to "hold high the banner of peace, development and cooperation" and be a responsible big country, and proposed to "create a harmonious world and seek mutual benefit and win-win." Under such circumstances, the security outlook of Chinese leaders has gone through significant change[70] and crisis management has gained increasing attention from the Chinese government.[71] Settlement of disputes through dialogue and avoidance of military conflict and war have become not only the main means to safeguard security but also an important objective in maintaining national, regional and world security. Crisis management has replaced military confrontation to become the main feature of China's crisis behavior.

It is noteworthy that against the above-mentioned macro background the constant strengthening and improvement of military and security crisis management mechanism, development of relevant principles, rising position in international regimes and progress in national defence modernization[72] have all contributed to enhancing China's capability of managing military and security crises and enabling it to take a more active part in certain regional or even global military and security crisis management. It is with the strong awareness and desire of crisis management and the markedly increased capability that the Chinese crisis behavior has undergone the big deep changes.

Future prospects

The various military and security crises confronting China have gone through and are undergoing profound changes. For a long time, China has to deal with mainly three types of military and security crises: military crisis, military conflict or war occurring outside of Chinese territory, in particular those happening in its surrounding regions; military crisis or contingency caused by territorial or maritime disputes between China and other countries; and military crisis caused by the Taiwan question. Since the end of the Cold War, particularly in the new century, all three types of crises have gone through profound changes.

First, military crises related to military conflicts or local wars caused by traditional security issues and geopolitics (especially between powers) have sharply decreased throughout the world, whereas crises and conflicts caused by non-traditional security issues have been on the rise. The influence of such a change is deep. Compared with traditional ones, in most non-traditional security crises, although various conflicts of interest and intense gaming continue, there is greater space for crisis management and international cooperation including among major powers, since states tend to have fairly large common interests.

Then, some disputes over territory or maritime rights and interests that caused military conflicts between China and other countries have been either fundamentally resolved or brought under control, without causing further military crises. Although as a contrast to the situation along land borders maritime disputes continue surfacing from time to time and even increasing, they are generally controllable. The main reasons for such a change are the improvement and development of overall political economic and security relations between disputing countries, the existence of various dialogue and negotiation mechanisms, the creation and strengthening of various military confidence-building measures and the gradual acceptance of the Chinese proposition of "putting aside differences and engaging in joint development" by relevant parties to the disputes.[73]

Finally, after recurring crises and constant tension for some time, a major turn has appeared in the situation across the Taiwan Strait, with cross-Strait relations showing a prospect of continued relaxation, improvement and growth. Although there is still a long way to go before the Taiwan question will be fundamentally resolved, striving for peaceful reunification has become an unswerving objective of China. Given that Chinese overall national strength will continue growing rapidly for a fairly long time to come and with the changing balance of political power within Taiwan, the probability of a crisis across Taiwan Strait has greatly decreased.

In this new situation, while further strengthening military crisis management, China will also make effort to strengthen conflict settlement and opportunity management and try to align the three. The three are closely linked and none is dispensable if China is to better safeguard its national security and promote regional and world peace and development.

Crisis management (including crisis avoidance and crisis control) is a process in which parties to a crisis both engage in a game and cooperate with each other. The main purpose is to strive for prevention of continued escalation of conflict of security interests ultimately into military conflict or war while trying one's best to maintain one's own interests. Crisis management remains the primary and most important path and means to deal with those major conflicts of interest that are unlikely to be resolved for a fairly long time in the future, such as certain Sino-foreign land and maritime territorial disputes, China–US maritime security frictions, traditional security crises that continue existing in the neighboring areas (including traditional Indo-Pakistani conflicts, South–North face-off on the Korean Peninsular and state of hostility between the DPRK and the United States) and some emerging non-traditional security crises (triggered by nuclear proliferation and international terrorism). In order to cope with these crises more effectively, China should make greater efforts in establishing bilateral and multilateral crisis management systems with pertinent countries.

Conflict settlement goes beyond temporary subsiding to remove the root causes of conflict, facilitate parties to the conflict to reconstruct their relations so that their objectives do not conflict with one another or enable them to realize that their objectives can be reached without conflict. It is therefore a more difficult, time-consuming and strenuous task.[74] If achieved, it will be of greater significance for peace and security. Conflict settlement will be the focus of attention by China in dealing with some disputes over maritime rights and interests and remaining differences across the Taiwan Strait. Besides, China will also work for conflict settlement once multilateral crisis management produces major results.[75]

Opportunity management requires the relevant states and parties to look beyond their differences of interest, utterly give up Cold War mentality, change the stereotype of proceeding from worst-case scenarios, seize all opportunities and conditions to actively develop any possible cooperation and strive for the best prospects in the security fields or on security issues where their interests converge. Such an effort will be irreplaceable for expanding common interests, increasing mutual trust and containing and defusing conflicts of interest. To deal with new international situation, new challenges and new relationships between powers, China, which is determined to follow a peaceful development path, must manage well all opportunities. The joint development and furtherance of crisis management, conflict settlement and opportunity management will become an outstanding feature of external relations of China.

As a matter of fact, in recent years, China has made great efforts in pursuing the foreign policy of ''building a harmonious world,'' actively advocated a new security concept that embodies comprehensive security, common security and cooperative security, initiated various new crisis management principles to pragmatic solutions to specific disputes, and formulated policies towards increasing non-traditional security challenges. All of these have identified the direction of merging crisis management, conflict settlement and opportunity management, which is now part of Chinese practice. Such an effort will accompany the whole process of China's peaceful development. Its success will not only forcefully promote China's own security and development but also make a major positive contribution to world peace, development and prosperity.

Notes

1 See Han Jiahe and Xiong Chunbao, ''Qiantan junshi weiji kongzhi [Perception on Controlling Military Crisis],'' *Zhongguo guofang bao* [*China's National Defence News*], October 22, 2001.

2 See Jonathan Wilkenfeld, ''Guoji weiji guanli yanjiu gainian he fangfa [Concepts and Methods in the Study of International Crisis Management],'' in Zhang Tuosheng and Michael D. Swaine (eds), *Zhongmei anquan weiji guanli anli fenxi* [*Managing Sino-American Security Crisis Management: Case Studies and Analysis*] (Beijing: World Knowledge Press, 2007), p. 83.

3 See also definition of international crisis in Zhao Pi (ed.) *Zhongguo junshi baike quanshu* [*Chinese Military Encyclopaedia,* 2nd Edition, Disciplinary Volume II, *International Military Security*] (Beijing: Encyclopaedia of China Publishing House, 2008), p. 476.

4 See Zhang and Swaine (eds), *Zhongmei anquan weiji guanli anli fenxi*, p. 91.

5 See Zhang and Swaine (eds), *Zhongmei anquan weiji guanli anli fenxi*, pp. 4–5.

6 The Taiwan Question is an internal affair of China. However, due to the long-term US policy of military intervention over Taiwan, this question has gained certain ''internationalness'' and the military crisis across the Taiwan Strait is usually manifested as a military crisis between China and the United States.

7 According to the Geneva Accords, France withdrew peacefully, Vietnam was separated into two zones by the seventeenth parallel and Laos and Cambodia would be neutral.

8 Besides public statements that China would not sit idly by while its neighbor was invaded, Chinese leaders also conveyed through the Indian ambassador to China a warning that the US troops should not cross the thirty-eighth parallel. See Kavalam Madhava Panikkar, *In Two Chinas: Memoirs of a Diplomat* (London: Allen and Unwin, 1955), p. 110.

9 The United States ignored the series of public warnings of China and misjudged that it was impossible for China to send troops to the war. See Allen S. Whiting, ''U.S.

Crisis Management Vis-à-vis China,'' in Michael Swaine, Tuosheng Zhang and Danielle Cohen (eds), *Managing Sino-American Crisis: Case Studies and Analysis* (Washington, DC: Carnegie Endowment for International Peace, 2007), p. 219.

10 Niu Jun, ''Chinese Decision Making in Three Military Actions Across the Taiwan Strait,'' in Swaine, Zhang and Cohen (eds), *Managing Sino-American Crisis*, pp. 309–310, 313.

11 For example, China in 1955 refrained from attacking the US and Chiang ships withdrawing from Dachen Island (before that the United States conveyed through the Soviet Union to China its hope that China would not attack the withdrawing ships) and in 1958 decided to ''strike only Chiang's ships but not US ships'' that transported supplies to Quemoy and Matsu.

12 With the two military campaigns in the Taiwan Strait, Chinese leaders came to realize that the Taiwan Question would be a long-term one due to US intervention. Mao Zedong decided to give up the original plan to capture Quemoy and Matsu so as to oppose the intention of the United States to create two Chinas by separating the two sides of the Taiwan Strait.

13 Disputed areas between the two countries include 90,000 square kilometers in the eastern section, 2,400 square kilometers in the middle section and 30,000 square kilometers in the western section.

14 See Xu Yan, ''Zhongyin bianjie chongtu shi shuangfang jundui suzhi, zhuangbei he zhanlue zhidao (Er) [Quality, Equipment and Strategy of Chinese and Indian Militaries During Border Conflicts],'' *Bingqi zhishi* [*Ordnance Knowledge*], No. 12B (2009).

15 At that time, China was in serious economic difficulty at home and Chiang Kai-shek clamored to ''re-conquer the mainland.'' Internationally, the United States remained hostile to China and relations with the Soviet Union started to worsen.

16 In the early 1960s, China provided moral and material support for the Vietnamese Communist Party in its armed struggle in the South. For US aircraft invading Chinese air space, China kept watch on their movements but refrained from attacking them. See Zhang Baijia, *Kangmei yuanchao yu kangmei yuanyue* [*Aiding Korea against the United States and Aiding Vietnam against the United States*], in Zhang and Swaine (eds), *Zhongmei anquan weiji guanli anli fenxi*, pp. 131–132.

17 See Zhang Baijia, ''China's Role in the Korea and Vietnam Wars,'' in Swaine, Zhang and Cohen (eds), *Managing Sino-American Crisis*, pp. 204–205.

18 See Xu Yan, ''1969nian de zhongsu bianjie chongtu [*Sino-Soviet Border Conflict in 1969*], *Dangshi yanjiu ziliao* [*Materials for CPC History Studies*], No. 5 (1994).

19 Starting from autumn 1978, Deng Xiaoping visited a number of Southeast Asian countries and went to the United States in January 1979, working to obtain their support or understanding of China's position of opposing Vietnamese expansion.

20 Vietnam started in 1988 to withdraw its troops from Cambodia and from the frontier of Sino-Vietnamese border. China thus stopped its cross-border operations.

21 See Yang Gongsu, *Cangsang jiushi nian, yige waijiao teshi de huiyi* [*Ninety Years of Vicissitudes, Memoirs of a Special Diplomatic Envoy*] (Haikou: Hainan Publishing House, 1999), pp. 281–284.

22 Actions taken by the Philippines included the demolishment of measurement marks erected on some islands and reefs by China, illegally seizing Chinese fishing boats and arresting Chinese fishermen and organizing foreign journalists to Meiji Jiao for ''reporting.'' See Li Jinming, ''Meiji jiao shijian de qianqian houhou [The Ins and Outs of the Meiji Jiao Incident],'' *Nanyang wenti yanjiu* [*Southeast Asian Studies*], No. 1 (2000), p. 68.

23 See *Renmin Ribao* [*People's Daily*], July 31, 1995.

24 In May 1995, the United States approved Lee Teng-hui to visit to the United States. In June, Lee delivered a political speech advocating Taiwan independence in Cornell University.

25 In August 1995, the United States informed China of the various restrictions over future visits to the United States by Taiwanese leaders; in 1998, President Clinton formally put forward the "three-no" policy towards Taiwan.
26 During the crisis, China all along maintained high-level contacts with the United States on international occasions and consultation between officials, among which the senior officials' consultation in Washington DC in March 1996 played an important role at the key moment.
27 The first aircraft carrier formation was deployed to the east of Taiwan and the second seemed to have deliberately delayed its time of arrival at West Pacific. Furthermore, all US ships and aircraft deployed in the region received instruction not to enter into the Taiwan Strait or intervene in the PLA exercises in any way. See Robert L. Suettinger, "U.S. 'Management' of Taiwan Strait 'Crises,'" in Swaine, Zhang and Cohen (eds), *Managing Sino-American Crisis*, pp. 278–279, 283.
28 For example, Lee Teng-hui in 1999 put forward the "special two states theory" and Chen Shui-bian in 2003 put out the theory of one state on each side of the Taiwan Strait and attempted on many occasions to decide the status of Taiwan through so-called "referendums."
29 On May 8, 1999, US missiles attacked the Chinese Embassy in the Federal Republic of Yugoslavia, leading to serious damage to the embassy premises and the deaths and injuries of over 20 Chinese. The United States claimed that it was bombed by mistake.
30 On April 1, 2001, a US surveillance aircraft off the southeast coast of Chinese Hainan Province collided with a Chinese military aircraft flying a monitoring mission. The Chinese aircraft crashed, its pilot missing; the seriously damaged American aircraft landed at an airport in Hainan.
31 The embassy bombing caused great indignation of the Chinese government and people. Mass protests and demonstrations against the United States broke out in many Chinese cities and the embassy and other diplomatic premises of the United States were once surrounded by angry public. The US consulate in Chengdu was set on fire.
32 For the embassy bombing, US public opinion generally agreed that the United States should take the responsibility and apologize to China. The aircraft collision, however, caused strong public reactions in both China and the United States and the confrontational sentiments grew rapidly.
33 Former Commander-in-Chief of US Pacific Command Admiral Blair once pointed out at a seminar that should the US side also lose its aircraft and crew (there were 24 on board), or should China decide to shoot down the US surveillance plane or should the US authorities approve the destruction of the US plane in Chinese territory by precision strike, the event might have very well evolved into a military conflict.
34 In the initial days of the incidents, normal contacts between the two countries failed to offer timely communication and the hotline between heads of states that had already been set up was not put to use.
35 China first released the US crew after the US side said that it was "very sorry" on the relevant issues. Then after negotiation, China allowed the United States to take the aircraft back using a civilian aircraft, after disassembly.
36 See *Tokyo Shimbun*, May 28, 2004; *Yomiuri Shimbun*, June 8, 2004.
37 The two countries then were in constant friction over the question of Diaoyu Dao and bilateral political relations were confronted with serious difficulties due to Prime Minister Koizumi paying tributes to the Yashukuni Shrine.
38 In May 1995, India suddenly conducted a nuclear test, followed by Pakistan. South Asia faced a dangerous nuclear arms race.
39 China presided over the meeting of P-5 Foreign Ministers, which adopted a joint communiqué, and facilitated the adoption of the UN Security Council resolution 1172, containing opposition to the nuclear tests by India and Pakistan, insistence on the nuclear non-proliferation regime, prevention of conflict and the nuclear arms

race between India and Pakistan, determination of not giving the two countries status of nuclear weapon states and decision to sanction the two states. See Tang Jiaxuan, *Jinyu xufeng* [*Storms and Breeze*] (Beijing: World Knowledge Press, 2009), pp. 406–408.

40 In October 2002, the United States accused the DPRK of secretly manufacturing enriched uranium in violation of the framework agreement, and the Second DPRK nuclear crisis broke out. After that, the DPRK withdrew from the Nuclear Nonproliferation Treaty and conducted two nuclear tests.

41 China supported the UN Security Council to adopt two resolutions sanctioning the DPRK for conducting nuclear and missile tests.

42 In early 2003, the Iranian nuclear issue became outstanding due to Iran's development of uranium enrichment capability and later evolved into another serious nuclear crisis.

43 China's position is similar to that of the United States, United Kingdom, France, Germany and other countries in opposing Iran acquiring nuclear weapons and demanding that it stop production of enriched uranium and to accept nuclear fuel exchange plans. However, on the question of sanctions, China stands for a more cautious attitude.

44 After 9/11, the United States waged an anti-terror war overthrowing the Taliban regime in Afghanistan. Then the United States decided to launch a military strike against Iraq on the grounds that the latter had been developing weapons of mass destruction.

45 At that time, the United States and United Kingdom tabled a one-step draft resolution containing automatic authorization of the use of force against Iraq by the United States, to which China, France and Russia objected. China presided over the adoption of a two-step UN Security Council Resolution, according to which large-scale unrestricted weapons inspection should first be carried out and the Security Council would make a decision on the next-step action after the inspection results are reported to it.

46 In May 1991, China and the Soviet Union signed an agreement on the eastern section of border, basically resolving the most controversial and most difficult part in their boundary disputes. After disintegration of the Soviet Union, China concluded border agreements with Russia, Kazakhstan, Kyrgyzstan and Tajikistan. By 2005 the boundary issue that originally existed between China and the Soviet Union and involved a 7,600 km border was completely resolved.

47 In December 1999, China and Vietnam formally signed the China–Vietnam Land Border Treaty. In December 2000, the two countries signed the Beibu Gulf Border Demarcation Agreement and Beibu Gulf Fishery Cooperation Agreement. Demarcation of the land border and boundary in the Beibu Gulf were properly settled.

48 In the past nearly 20 years, the Chinese and Indian governments have engaged in consultation and negotiation to strive for an early resolution of their border disputes and established some important military confidence-building measures, thus maintaining peace and stability in the disputed areas.

49 The then Foreign Minister Qian Qichen shuttled countries in the Middle East including Iraq; Chinese and American leaders had consultation in various forms; in the end China abstained when the UN Security Council voted a draft resolution allowing military operations against Iraq. See Qian Qichen, "Fei wang bageda [Flying to Bagdad]," in *Waijiao shiji* [*Ten Stories of a Diplomat*] (Beijing: World Knowledge Press, 2003).

50 See Wang Xiaodong, *Guojia anquan lingdao tizhi yanjiu* [*A Research on National Security Leadership Structure*] (Beijing: Shi Shi Publishing House, 2009), p. 324.

51 In the first 40 years of the PRC, leaders such as Mao Zedong, Zhou Enlai and Deng Xiaoping had decisive influence over decision-making. As professional revolutionaries and leaders for life, their experience, wisdom, leadership and high prestige left deep imprints on Chinese security decision-making, including crisis behavior, with both successful experiences and profound lessons.

52 The Central National Security Leading Small Group was formed after the embassy bombing incident in 1999. It played an important role in handling the aircraft collision incident. The forming of this small group was an important attempt of China to strengthen the military and security crisis management.

53 These central leading small groups are composed of top leaders and leaders of relevant functional departments. Their compositions are both different and overlapping. Such a mechanism has the benefit of clear definition of responsibility and focus but it is weak in thinking through the various security/military crises and in having unified decision-making and coordination.

54 This has been driven by enhanced awareness of the importance of information and intelligence (including hard intelligence and soft intelligence) and great improvement in relevant technical means.

55 Many Chinese scholars hold the view that the guidelines of "observe calmly; secure our position; cope with affairs calmly; hide our capacities and bide our time; be good at maintaining a low profile; never claim leadership and make some contributions" put forward by Deng Xiaoping also played a role in guiding military and security crisis management since the 1990s.

56 For example, the China Foundation for International Strategic Studies and the Carnegie Endowment for International Peace jointly put forward eight basic principles for China and the United States to conduct crisis management: 1, Maintain direct channels of communication and send signals that are clear, specific and detailed; 2, Focus on limited objectives and employ means on behalf of such objectives; sacrifice unlimited goals; 3, Preserve military flexibility and civilian control, escalate slowly and respond symmetrically; 4, Avoid ideological or principled lock-in positions that encourage zero-sum approaches to a crisis and limit options or bargaining room; do not confuse moral or principled positions with conflicts of interest; 5, Exercise self-restraint, and do not respond to all provocative moves; 6, Avoid extreme pressure, ultimatums or threats to the adversary's core values; and preserve the adversary's option to back down in a "face saving" manner; 7, Divide large, integrated, hard-to-resolve disputes into smaller, more manageable issues, thereby building trust and facilitating tradeoffs; 8, Think ahead about the unintended consequences of one's actions.

57 The desire to cooperate and necessary compromise are always necessary conditions for crisis management to be successful.

58 Referring here mainly to the behavioral paradigm in managing military and security crises between China and other countries rather than those between other countries.

59 Including public diplomatic declarations, statements of leaders and messages convey through third parties.

60 Such as moving troops, engaging in limited operation and conducting military exercises.

61 That means always putting the maintenance of international justice and national dignity at very important positions.

62 See Wu Baiyi, *Zhongguo guoji weiji guanli gailun* [*Introduction of Chinese International Crisis Management*] (Ph.D. Dissertation, Graduate School at Chinese Academy of Social Sciences, 2010), pp. 76–78.

63 See Zhang Tuosheng, "On China's Concept of the International Security Order," Robert S. Ross, Øystein Tunsjø and Zhang Tuosheng (eds), *US–China–EU Relations: Managing the New World Order* (London: Routledge, 2010).

64 China regained its legitimate seat in the United Nations in 1971. However, for quite some time since then, China remained outside of many major international organizations.

65 On crisis, Mao Zedong stressed more the aspect of "bad things turning into good," "chaos leading to great order," and "war causing revolution and revolution stopping war," but failed to fully appreciate the dangers of crisis per se. As Chinese scholar

Niu Jun pointed out, Mao Zedong's sense of crisis was mainly threefold. First, he believed that tension and conflict were the most basic features of world politics. Second, he was especially sensitive to threats. And third, he was anxious to pursue radical changes or leap-frog progress. See Niu Jun, "Mao Zedong weiji yishi yu zhongsu tongmeng puolie de yuanqi [Mao's Crisis Perception and the Beginning of the Sino-Soviet Schism]," in Zhang Baijia and Niu Jun (eds), *Lengzhan yu Zhongguo* [*The Cold War and China*] (Beijing: Shijie zhishi chubanshe, 2002), p. 275.

66 Winning or losing a war no long determines everything and the purposes of war are increasingly limited. The political gains and losses and whether the results would be beneficial to economic development "have become the most important social criteria to judge the success or failure of a war in the time of economic globalization." See Guo Weitao and Zhu Shaopeng, *Zhanzheng, zhanlue, jundui* [*War, Strategy and Military Forces*] (Beijing: Military Science Publishing House, 2010).

67 Big countries cooperate and compete with one another at the same time and in general their relationships are no longer confrontational in nature.

68 An important decision made by Deng Xiaoping was to not draw a line with ideology nor support "armed struggles" within any surrounding country.

69 The new security concept stresses comprehensive security, common security and cooperative security and regards economic cooperation as an important basis to strengthen security cooperation.

70 Now Chinese leaders observe various crises mainly from the perspective of the possible large negative influence on the peaceful development of the region and the world and on China's modernization drive. Peace, stability and harmony have become the main starting point of Chinese foreign policy.

71 Chinese leaders began to pay greater attention to security crisis management at the turn of the century. In the face of the various Chinese–foreign and international military and security crises, a new view gradually came into being: in the new world situation, various security crises will replace direct war to be the primary challenge and it is imperative for the state to improve the crisis management mechanism and strengthen crisis management capability.

72 Effective and reliable military deterrence capability is an indispensable condition and powerful support of successful crisis management.

73 Although no substantive progress has been made up to now in joint development, it remains the correct path to resolution of relevant disputes.

74 See Jonathan Wilkenfeld, "Guoji weiji guanli yanjiu gainian he fangfa," pp. 99–100.

75 For example, the Six-Party Talk on the DPRK nuclear issue has the crisis management function, but its main purpose is to resolve the relevant conflict. Only when the DPRK gives up its nuclear weapon program, security assurance is provided to the DPRK, and a peace mechanism is put in place on the Korean Peninsula, can long-term peace be realized on the Peninsula.

4 Managing ethnic minority crises

The Tibetan areas and Xinjiang

Colin Mackerras

This chapter aims to analyze China's crisis management among its ethnic minorities through an exploration of four events or series of events in the Tibetan areas and Xinjiang. These four constitute the most serious crises to have taken place in China's ethnic areas since the Cultural Revolution and even before. The time scope begins with the independence demonstrations in the Tibetan capital Lhasa in the autumn of 1987 and ends in 2010 with the aftermath of the Tibetan 2008 disturbances and the July 2009 riots in the Xinjiang capital Ürümqi. For Xinjiang the ethnic minority with the greatest focus is the Uygurs, a Turkic people with great dedication to Islam, while for the Tibetan areas it will obviously be the Tibetans, who adhere strongly to their own form of Buddhism.

The chapter has three broad sections. The first defines what constitutes a crisis in the Tibetan areas and Xinjiang in the context of China's imperatives there. It also contains some essential preliminary material, such as background information on the minorities, some precise analysis of the territories under consideration and a very brief description of the four crises being discussed. Finally, it notes the structure of crisis management in the relevant areas. The second and third sections are comparative analyses respectively of the internal and external management of the selected crises, taking account both of immediate and longer-term management. The conclusion makes some evaluation of China's crisis management in the Tibetan areas and Xinjiang and argues that, despite some weaknesses, it is effective enough to the render the break-up of China improbable within the foreseeable future.

Introduction and background material

The 2000 census showed China's 55 officially recognized minority nationalities (*shaoshu minzu*) at 8.41 per cent of the total national population of 1.26583 billion, accounting for 106.43 million people.[1] The less definitely accurate but more recent figures from the National Sample Survey held in 2005 put the minorities at 9.44 per cent of the total of 1.30628 billion.[2] Neither figure included Hong Kong, Macau or Taiwan.

Defining crises: China's imperatives in the Tibetan areas and Xinjiang

The crises considered in this chapter are an important subcategory of the "collective protests" that are the core of Jae Ho Chung's contribution (Chapter 2 in this volume) and the definition of "crisis" follows his. Applied to the ethnic minority areas, a crisis would fall into one or more of the following categories:

- an independence or quasi-independence movement that flares into violence or big demonstrations;
- serious ethnic rioting, with members of one ethnic group turning on another;
- serious collective protests in ethnic areas perceived as threatening social, political or economic stability.

The features of crises as defined here flow from China's top imperatives in the Tibetan areas and Xinjiang. The first is to remain united, which means opposing any attempt to separate these western areas from China. The preamble of China's Constitution specifies that the People's Republic is "a unitary multinational state built up jointly by the people of all its nationalities." Article 4 prohibits any act that undermines "the unity of the nationalities" or instigates secession. It declares that all areas where ethnic minorities exercise autonomy are "inalienable parts of the People's Republic of China."[3] The *Law of the People's Republic of China on Regional National Autonomy* contains very similar provisions.[4]

The desire to remain united implies strong opposition to separatism or to independence for any of China's territories. It follows that the break-up of the Soviet Union at the end of 1991 was of great concern to the Chinese leadership, especially since several of the newly independent states border China. However, China's leaders did not relax their requirement for national unity, and it anything intensified it. Authorities believed that if border territories like Tibet and Xinjiang succeeded in moving away from Beijing's control, the break-up of China would become a realistic threat.[5]

The second main Chinese government imperative for the Tibetan areas and Xinjiang is stability of various kinds, which is very similar to internal security. The avoidance of ethnic rioting ranks very high in the wish to maintain stability. Because Jae Ho Chung has considered the issue of political stability in his chapter in this volume, I leave aside further discussion here.

The other side of the coin of maintaining stability is the goal of establishing a harmonious society (*hexie shehui*), which the Hu-Wen leadership introduced in 2004 for CCP (Chinese Communist Party) rule. In a speech of February 19, 2005, Hu Jintao explained that the harmonious society would feature democracy, the rule of law, equity, justice, amity and vitality, among other desirable qualities. He said it would improve relations between the people and government, adding that it "will result in lasting stability and unity."[6] The converse, to be avoided at all costs, would be "disorder" (*luan*), which is exemplified in ethnic riots and rebellion.

The third imperative is economic development. The government is convinced that, if it can ensure a high standard of living among the ethnic minorities, it will be able to persuade them that remaining within the People's Republic is in their best interests. In 1999, the government decided on a campaign "to open up the west" (*xibu da kaifa*), this area including all the Tibetan areas and Xinjiang. In his report to the National People's Congress on March 5, 2000, just after the campaign had actually started in January 2000, Premier Zhu Rongji stated that, apart from promoting sustained economic growth and boosting domestic demand, opening up the west would "strengthen national unity, safeguard social stability and consolidate border defense."[7] As one authority points out, these are "barely coded phrases for being concerned about issues surrounding" the ethnic minorities.[8]

Policy

We turn now to a brief summary of China's policy on its ethnic minorities. The essence of policy is that all ethnic groups are equal, and minorities may practise autonomy within designated areas, but since territorial unity is China's top imperative, secessionism is absolutely forbidden. Among the 55 ethnic minorities the Tibetans and Uygurs have posed by far the most serious problems to the government in terms of separatism.[9]

The four constitutions of the People's Republic of China (PRC), and most notably that of 1982, form the legal basis of autonomy. There is more detail in the *Law of the People's Republic of China on Regional National Autonomy*, adopted in May 1984 and amended in February 2001. The categories of autonomous places are autonomous regions, which are equivalent to provinces in status, autonomous prefectures (AP) and autonomous counties (AC). Two of the five autonomous regions are relevant to this chapter: the Xinjiang Uygur Autonomous Region (XUAR) and the Tibet Autonomous Region (TAR).

One of the main principles of autonomy is to train and promote cadres from the ethnic minorities. Also, top government leaders of the autonomous places must belong to the ethnic group exercising autonomy.[10] However, a major limitation is that there is no such requirement for leaders of the CCP, which is actually much more powerful in China than the government. As we shall see later in this chapter, decisions made by Han Party secretaries can overlook ethnic requirements as against those of the state.

Despite limitations such as this one, the tone of the autonomy laws is generally quite liberal. Principles include respect for local cultures and freedom to use local languages, including in government, and encouraging Han cadres to learn local languages.[11] Article 11 of the law, as amended in 2001, lays down freedom of religion in terms identical with the Constitution (Article 36 of the 1982 version still current),[12] including a ban on foreign domination.

Territory under consideration

Although the minorities constitute only a small proportion of the total population, they take up about two-thirds of the territory recognized internationally as China's. Most of the minority areas are in the geographical western half of China, which is the subject of this chapter. Xinjiang is relatively simple. It is China's north-westernmost province-level unit, and takes up about one-sixth of China's total territory.

The "Tibetan areas" are much less straightforward. The Chinese government considers these areas to mean the TAR, ten autonomous prefectures where it regards the Tibetans as exercising autonomy, in two of them along with another ethnic group; and two autonomous counties. Of the ten autonomous prefectures, two are in Sichuan Province, one in each of Gansu and Yunnan, while the remaining six are in Qinghai. Of the two autonomous counties one is in Sichuan, the other in Gansu. The six autonomous prefectures in Qinghai take up the whole of the province, apart from the provincial capital Xining and Haidong to its east. An additional complicating factor is that the Tibetan Government-in-Exile (TGIE), of which the fourteenth Dalai Lama is the leading figure, regards "Tibet" as meaning not only the TAR, but all of Qinghai, and the current autonomous prefectures and counties in Sichuan, Yunnan and Gansu.[13] In other words, the TGIE's concept of "Tibet" means a territory slightly greater than the "Tibetan areas" as the Chinese government understands them, because it includes the whole of Qinghai, even the far eastern segment where Xining and Haidong are located. Either way, the area is about 2 million square kilometres, and just under one-quarter of China's total territory.

According to the 2000 census, there were 8,399,393 Uygurs in China, of whom 8,345,622 or over 99 per cent, lived in Xinjiang, and 5,416,021 Tibetans, of whom just over 5 million (about 92 per cent) lived in the Tibetan areas. Uygurs were 40.6 per cent of Xinjiang's total, while Tibetans made up 68.67 per cent of the population of the Chinese-defined Tibetan areas.[14] For comparison, in the 1990 census, there were about 7.2 million Uygurs in Xinjiang, making up 47.5 per cent of the region's total population of 15.2 million, and 4.6 million Tibetans living in the Tibetan areas, about 78.7 per cent of the total population of 5.845 million.[15] The main reason for the falls in the Uygur and Tibetan proportion of the population in their respective areas is Han immigration. This factor causes resentment to the Uygur and Tibetan populations.

The four crises

To analyze crisis management necessitates some basic information on the relevant incidents. This chapter focuses on four crises, all of them fitting the definition offered in the subsection "Defining crises." These are (i) the independence demonstrations and disturbances in Lhasa and other Tibetan areas from 1987 to 1989; (ii) the Baren Township uprising in Xinjiang in 1990; (iii) the Tibetan area disturbances of March and April 2008; and (iv) the Ürümqi riots of 2009.

There were three major demonstrations in Lhasa on September 27 and October 1 and 6, 1987, mainly by monks demanding independence, of which those in October turned violent. In 1988 disturbances took place in several Tibetan areas, reaching a climax on March 5, when thousands of Tibetans demonstrated in Lhasa for independence on the last day of a major traditional Tibetan ceremonial period. From March 5 to 7, 1989 further independence demonstrations took place in Lhasa, aimed at commemorating the thirtieth anniversary of the 1959 rebellion that had led to the Dalai Lama's flight to India. These developed into full-scale riots, with looting of Han houses and shops and quite a few casualties. Meanwhile, the Tibet issue became strongly internationalized. In March 1988, very shortly after the March 5 demonstration, the International Campaign for Tibet was established. At the end of 1989, the Dalai Lama was awarded the Nobel Prize for Peace.

Very soon afterwards a crisis erupted in Xinjiang. What happened, according to a report carried on Xinjiang television on April 21, 1990, was that on April 5 some counter-revolutionaries in Baren Township, Akto County, not far from Kashgar in the southwest of Xinjiang, attacked security personnel in a local government building, killing six people. The next day a counter-attack by the local authorities left 13 people dead. The "counter-revolutionary rebellion" was soon suppressed and order restored.[16]

What caused this incident? Local reports said that closing a mosque before a religious festival had led to gatherings in village mosques demanding more freedom for Islam on April 4, the riot the next day involving about 2,000 villagers, almost all local Uygur, Kirgiz and Hui peasants.[17] Much more important, however, are the larger and longer-term causes. Sources both opposing and supporting this uprising suggested that this was a fairly deliberate, if not well planned, attempt to stir up a *jihad* (holy war) with the aim of getting rid of the Chinese and setting up a state independent of China.[18]

The 1990s and first years of the twenty-first century were comparatively free of problems in the Tibetan areas in the sense that there was no repeat of demonstrations on the scale that had marked the late 1980s. The reasons for this relative calmness included tighter security in the Tibetan areas, good economic development and the failure of the external Tibet lobby to find resonances within China itself. In Xinjiang, on the other hand, the situation continued very unstable for the 1990s, and there were quite a few very serious incidents. On the other hand, the first few years of the twenty-first century appeared to be reasonably stable.

The apparent calm was abruptly broken in 2008. Following a demonstration by monks on March 10, serious ethnic rioting mainly by lay Tibetans against Han and Hui flared in a violent incident in Lhasa on March 14, with casualties and serious destruction of shops and other property. Slogans included "independence for Tibet" and support for the Dalai Lama. On following days, numerous incidents erupted all over the Tibetan areas, at least 15 of them involving violence against police stations and other state property.[19] Robert Barnett's analysis shows that 74 out of 95 incidents reported as taking place from March 10 to April 5, 2008 were outside the TAR.[20] Whereas most participants in past disturbances had been monks, in 2008 the majority were laypeople. Casualty reports fluctuated wildly. Xinhua claimed that 19 people were killed in Lhasa on March 14, with isolated others in the Tibetan areas of Sichuan and Gansu, mostly Han.[21] However, the Dalai Lama's Central Tibetan Administration put the number of dead for all Tibetan areas over the period March 10 to April 25, 2008 at 203. Causes were both internal and external, including serious and growing inequalities and greater effectiveness of the Tibet lobby in the West, especially giving the Dalai Lama the Congressional Medal of Honor late in 2007 and deliberate attempts to put the spotlight on China for human rights abuses in the lead up to the Olympic Games, to be held in Beijing in August 2008.[22]

The last of the four crises considered here was again in Xinjiang. On July 5, 2009 a protest demonstration by Uygurs in Ürümqi developed into the worst ethnic rioting in the PRC's history, with Uygur mobs turning against Han and Hui. On July 7, Han counter-attacked against Uygurs. According to official Chinese figures, the final death toll was 197, with over 1,600 injured. Among the dead, 156 were termed "civilians," and among these 134 were Han, 11 Hui, 10 Uygurs and 1 Manchu.[23] There was also great damage to property. The following month, local Han blamed a series of attacks with hypodermic syringes on Uygurs. Xinjiang television reported that by September 2, hospitals had reported treating 476 people, of whom 433 were Han. However, despite rumors that the syringes were contaminated, police said nobody had caught any infections.[24] The initial spark that set this fire alight was an incident in a toy factory in Shaoguan, Guangdong Province, in which Han had murdered two Uygurs rumored to have raped Chinese women. However, the causes lay much deeper, and included deep Uygur resentments against the Han and the Chinese Han-dominated state, serious economic inequalities and both internal and external fanning of hatreds.

The structure of crisis management in the Tibetan areas and Xinjiang

Before detailed analysis of how the Chinese state responded to these four series of events, it is necessary to consider the mode or structure of crisis management in the relevant areas. This treatment of structure considers those bodies that take overarching decisions on how to respond to crises, and those that actually undertake the task of restoring order and stability in cases where these have been threatened or collapsed. Most of the arms of this structure are designed to cope with crises all over the country, but a few are specific to the minority autonomy places. The bulk of the structure predates the earliest of the crises under discussion here, but a part is somewhat more recent and may have been established to some extent because the crises occurred in the first place.

We begin with the structure's internal sector. The seriousness of the crises in the Tibetan areas and Xinjiang has required the involvement of the country's most powerful leadership group, namely the Standing Committee of the Politburo, government ministries or work conferences organized by the CCP Central Committee and the State Council. By the nature of its work the Ministry of Public Security looms large in handling crises involving large public demonstrations, riots or disturbances. Jae Ho Chung's chapter has considered bodies such as the Stability Preservation Leading Group, at the head of which is Zhou Yongkang, who in the twenty-first century has emerged as possibly the most powerful of all Chinese leaders in the specific field of security. Minister of Public Security from 2002 to 2007, in which role he carried out reforms aimed at making the police more professional, Zhou then joined the Standing Committee of the Politburo with ultimate responsibility for state security.

Regional crisis management has seen the close involvement of regional CCP secretaries. Although it goes without saying that the CCP is thickly involved in all activities in China, including crisis management, party secretaries like Wang Lequan and Zhang Qingli, took a particularly close interest in security. Wang Lequan was the Xinjiang CCP secretary from 1994 to 2010, acquiring the nickname "the stability secretary" for "his ability to step into chaos and haul it to order."[25] Zhang Qingli had been Wang's deputy in Xinjiang until 2006, when he became CCP secretary in Tibet. With views on security similar to Wang's he took direct control of the 110 Command Centre, established in Lhasa "with the primary objective of suppressing the [2008] disturbances and restoring full central government control."[26]

Important for Tibet specifically are the Forums on the Work in Tibet, of which there have so far been five, the first in 1980, the fifth in 2010. Although they are not specifically aimed at solving crises but at considering ways of improving the situation in Tibet generally, they can be part of the structure that manages crisis in Tibet. There will be more to say below on the Fifth Tibet Work Forum, the decisions of which were aimed at reducing tensions following the 2008 disturbances in the Tibetan areas. Such work forums for specific areas usually take place in Beijing, but a counter-example is the May 1996 work conference, which took major decisions on the situation in Xinjiang. Though held at the behest of the CCP Central Committee, it was organized by the Xinjiang CCP and held in Ürümqi.

On the external level, crisis has on occasion been a driving force for participating in international bodies and even taking the initiative for establishing them. The best illustrative example was in April 1996, just before the beginning of the Ürümqi meeting discussed in the preceding paragraph, when the Presidents of China, Russia, Kyrgyzstan, Kazakhstan and Tajikistan met in Shanghai and signed the *Treaty on Deepening Military Trust in Border Regions*. By this action they formed the Shanghai Five. In June 2001, this body increased its size and became the Shanghai Cooperation Organization. The Shanghai Five initially aimed "to establish a coherent security and foreign policy framework" in the Central Asian region and to demilitarize the border regions with what had been the Soviet Union.[27] However, all involved countries also had their own interests in combating terrorism and separatism. China was the only country of the five that had not been part of the Soviet Union.

At the level of enforcing stability when crises erupt, the most important body is the People's Liberation Army (PLA), which can be used both to maintain peace and to crush the most serious cases of rebellion, ethnic rioting or large-scale demonstrations pushing for independence of one or more of China's ethnic territories. One body involved in maintaining stability all over China, including the suppression of ethnic unrest, is the People's Armed Police (PAP). Dating from the beginning of the PRC, it was incorporated into the military in 1950 as the PLA Security Army. In 1983 the PAP was formally inaugurated as a body separate from the PLA, control being in the hands of the Central Military Commission and the State Council, but the Ministry of Public Security was responsible for overseeing operational activities.[28] Its missions are to maintain domestic stability and the CCP's hold on power.[29] The PAP has been actively involved both in Tibet and Xinjiang. Tai Ming Cheung also claims that in July 1995, several PAP regiments from northern Xinjiang took part in a 5,000-km exercise that lasted 20 days aimed at dealing with Uygur unrest in the south.[30]

Internal management of crises from 1987 to 2010

Since the four crises under consideration were selected because they were the most severe in China's ethnic areas, some discussion of how they were individually handled is necessary. However, there is also reference to similarities and differences in this management.

Short-term response

The first and most obvious method of short-term crisis management is police suppression. In 1987 the demonstrations of October 1 and 6 saw police firing at crowds, with casualties resulting. On March 5, 1988 PAP units entered the Jokhang Monastery in Central Lhasa and beat up monks who had fled there for safety, killing some and arresting others.[31] Police broke up the main demonstration with electric cattle prods and tear gas, with many arrests and widespread torture of prisoners.[32] The third stage came in March 1989, with "massive display of military force in the streets of Lhasa."[33] Although this repression did succeed in suppressing separatism, it also "sharpened Tibetan political consciousness and enlarged Tibetan demands for independence to encompass a range of modern political ideas,"[34] meaning that even a quiet situation might hide potential trouble for the future.

The Baren Township uprising of April 1990 also saw the authorities mobilizing troops and the PAP, which quickly put the disturbance down. Over the next few weeks, they rounded up more than 1,000 suspects across Xinjiang. There was a lull after that, but further incidents led to increased but counter-productive surveillance and repression.[35]

Despite the similar pattern of treatment, there were differences. The disturbances of March 1989 caused the Chinese authorities to impose martial law and it was not lifted until April 1990. This was the first time in PRC history that the authorities had imposed martial law, a strategy it was to repeat in Beijing two months later in response to the student demonstrations there. It seems that suppression of the Tibetan independence demonstrations loomed very large in the Chinese leadership's priorities. Yet there was no obvious internal leader of these demonstrations. Such was not the case with the Baren Township uprising. Its leader was Zeydin Yusuf, whom Chinese forces killed in suppressing the uprising. Moreover, his ideology of Islamic *jihad*, killing the Han infidel and overthrowing the state[36] was much more dangerous than the Tibetan Buddhism that underlay the Lhasa disturbances of 1987 to 1989.

Moving forward to the 2008 and 2009 incidents in Lhasa and Ürümqi respectively, we still find police reaction, but apparently more reluctant than for the earlier crises. In Lhasa, the PAP were initially very restrained in dealing with the monks' demonstrations on March 10 and later, but were caught off-guard by the intensity of the rioting on March 14. They needed orders to shoot at rioters, and were slow to bring them under control, by which time much of the damage had been done. Additional police and troops were brought into the Tibetan areas from outside, once the extent of the disorder became clear.

When the July 5, 2009 riots erupted in Ürümqi, the authorities sent in PAP and other security forces to control the situation. However, television footage at the time showed armed young men, mostly Han, in very restrained poses as local Uygurs verbally abused them. Moreover, the rules of engagement required authorization from the central authorities. As in Lhasa the preceding year, they were taken by surprise and quite slow to act. The fact that the riots took place on a Sunday evening also contributed to delay in government response. However, finally they did stop the violence and restrained the Han counter-attacks of July 7. Official reports said that security forces shot and killed 12 people engaged in violent or criminal acts.[37] Repression became even more severe after the riots than the already serious extent that had obtained before.[38]

We might note some other differences in short-term reaction between the 2008 and 2009 incidents. One of particular significance was Hu Jintao's personal reaction, which was low-key in the former case, but far higher in profile for the latter. When the July 5 riots occurred, he was in Italy attending a summit of the Group of Eight (G-8). Despite the importance of this meeting, he cut the visit short and returned to China, later visiting Xinjiang. He obviously considered that handling the crisis demanded his personal presence.

Another difference was reaction to the foreign media and outside world. The authorities' reaction in March 2008 was to expel the foreign media from Lhasa and the Tibetan areas. Later in the month, central authorities invited one highly select group to visit Lhasa. However, when they visited Central Lhasa's Jokhang Monastery, several monks shouted that they wanted freedom and questioned government reports that the Dalai Lama had been responsible for the disturbances.[39] The Chinese government responded by banning Western media entry into the Tibetan areas. In the case of Xinjiang, there was no total ban on the foreign media, suggesting that the Chinese authorities had changed, perhaps improved, their policy on this matter. On the other hand, communication with the outside world was more tightly controlled and censored. For nearly a full year, the Internet was cut off, making email contact impossible.

Longer-term management

There were strong parallels in the medium- to long-term management of the crises in the Tibetan areas and Xinjiang. All four crises considered here resulted in meetings on policy, with consistent calls for more emphasis on economic development, along with zero tolerance for any separatist activities. The campaign "to open up the west" affected both regions, aiming at promoting outside investment and economic growth. In both places there were implications for the regional leadership, although the specific modes, timing and choices of appointments showed differences over the four crises. All four crises resulted in harsh punishment of those found guilty of stoking or taking part in rioting.

There were also differences. "Patriotic education" campaigns aiming to counter separatism were a common feature of life in the monasteries of the Tibetan areas from the mid-1990s. "Strike hard" campaigns try to suppress crime generally in China, but in Xinjiang they apply to attempts to suppress Uygur separatism and "illegal religious activities," including ensuring "that imams do not teach Islam to children, do not advocate Islamic 'fundamentalism' or radicalism (however the state may define those terms), and do not encourage or create connections between Muslims in China and elsewhere."[40] Though these two types of movement were similar to each other, treatment of Islam was more iron-fisted than of Tibetan Buddhism.

Policy decisions

Crisis has usually led to rethinking and/or adjusting policy towards the Tibetan areas and Xinjiang. The mid-1980s saw a Tibetanization policy that promoted appreciation of Tibetan culture and more tolerance of Tibetan Buddhism. But the late 1980s crisis resulted in a change towards a more rigid approach that put slighter weight on Tibetan culture and more on rooting out any separatist tendencies, buttressed by patriotic education campaigns, and promoting economic growth and stability. This change began at a meeting of the Politburo in November 1987, just after the first demonstrations, but gathered momentum over the next months and years. The demonstrations appeared to vindicate the hard line, because it was while the Tibetanization policy was in force that the demonstrations had erupted.[41] The theme of the Third Forum on the Work in Tibet (July 1994) was "with one hand grasp development, with the other hand grasp stability." This referred to a binary between expanding the economy and improving the people's standard of living on the one hand and maintaining social stability by preventing separatism on the other.[42]

The follow-on from the Baren Township uprising in Xinjiang was quite similar and included the binary noted in the preceding paragraph. The Xinjiang Regional CCP held a meeting in July 1990 which reached the conclusion that the authorities had not devoted nearly enough attention to combating separatism: "The task of fighting ethnic separatism has not been carried out in a top-to-bottom fashion. This lesson is very important, and we still need to pay full attention to it today."[43] The final communiqué of the May 1996 work conference in Ürümqi declared faster economic development and raising the people's livelihood to be "the most important, basic work to do to ensure Xinjiang's stability."[44] It said that only by narrowing the differences between the coastal and western areas, accelerating economic development and raising the standard of living would it be possible to convince people to support the CCP and socialism and prevent separatist activities.

Xinjiang policy after the Baren Township uprising particularly targeted Islam because its ideological base had been the doctrine of *jihad*. Stricter surveillance was implemented to control Islam and Uygur and other Muslim professionals and cadres received warnings that they were in danger of losing their jobs if they continued attending the mosque. Authorities screened the Islamic clergy to weed out those they believed untrustworthy and likely to stir up *jihad*.[45] In September 1993, the Xinjiang People's Congress introduced a regulation, which is found only in Xinjiang, prohibiting parents from allowing minors to participate in religious activities.[46] In line with this regulation, Muslim schools were also closed down, while the number of theological training schools in Xinjiang dropped. The Ürümqi May 1996 work conference charged that some CCP committees were dominated by Muslims and demanded their reorganization, as well as better training for cadres,[47] since these are potential members of the committees.

Policy following the 2008 Tibetan disturbances and 2009 Ürümqi riots followed mutually comparable patterns, with only non-essential differences in emphasis. One important point of similarity was the need to recruit far more members of the minorities to the PLA, police and similar bodies. Authorities had been unable to communicate with the people adequately, and even failed to understand text messages sent to minorities planning the riots, apparently due to total lack of knowledge of the local language and conditions. Both ethnic sensitivities and practicalities demanded far more Tibetans and Uygurs in the security forces.[48]

After the Tibetan disturbances, the authorities initiated another "patriotic education" campaign. This applied not only to Tibetan Buddhist monasteries throughout the Tibetan areas, but also to laypeople and officials. For example, on April 21, the official *Tibet Daily* reported an education drive directed mainly at Tibetan officials and CCP members that aimed to "fight separatism, protect stability and promote development."[49]

The leaders also gave further thought to policy. At government request, a consortium of lawyers from Peking University advised on current policy on nationalities, and especially on the Tibetans. However, when the Chinese government issued its next white paper on ethnic minorities on September 27, 2009, it showed very little policy change. Most of the paper was about conditions among the minorities and restated the traditional autonomy policy in very positive terms. While not referring specifically to recent disturbances or riots, it did mention the defeat of all "acts aimed at splitting the country, including plots for the 'independence of Tibet.'" In an excessively positive overall evaluation of the situation in China it affirmed: "Today, the Chinese nation has become a name with which all ethnic groups in China identify themselves and to which they give their allegiance."[50]

The year 2010 saw two high-level and important policy meetings on ethnic issues. The first was the Fifth Tibet Work Forum, which took place from January 18 to 20 and covered all Tibetan areas, not just the TAR. The outcome was a significant boost for development and investment. According to CCP General Secretary Hu Jintao, it was essential to narrow economic gaps, especially those between the rural areas and the national average. He said that "more central government funding would go to improving livelihoods and social welfare, and boosting agriculture and animal husbandry," with less focus on infrastructure.[51] The Forum also introduced a dictum Hu Jintao called "going down the road of development with Chinese characteristics and Tibetan flavour" (*Zhongguo tese, Xizang tedian*),[52] which suggested more emphasis on Tibetan features than had previously been obvious. On the other hand, Hu Jintao spoke of "earnestly maintaining the normal order of Tibetan Buddhism, and guiding Tibetan Buddhism to keep in line with the socialist society," appearing to offer no relaxation on preventing any hint that religion might be used for anti-state purposes.[53]

The Xinjiang counterpart meeting followed in May 2010, the central government holding a special three-day top-level working conference, claimed as the first of its kind since 1949. On May 20, Hu Jintao announced that the conference had approved a "massive support package that includes preferential tax and resource policies," in the hope of building a "moderately prosperous society in all aspects" in Xinjiang by 2020. Hu said that Xinjiang "should improve people's living standards and build an eco-friendly environment, as well as ensure ethnic unity, social stability and security."[54] According to Ashar Turson, President of the Xinjiang University of Finance and Economics, the change in tax policy would increase Xinjiang's annual fiscal income by 8 to 10 billion yuan, a major boost for the region, since the 2009 figure was 38.8 billion yuan, and that the additional income deriving from the new policy would go mostly to poor areas. He said also that "economic development and improving people's livelihood are key to maintaining the region's long-term stability, because people are unlikely to be manipulated by terrorists or separatists when they are happy with their lives."[55]

Leadership ramifications

Both in Tibet and Xinjiang, the disturbances and riots led on to changes in leadership, usually related to the changes in policy outlined in the subsection "Policy." While leadership changes thus tend to *follow* policy shifts, which themselves often result from crises, they are still part of crisis management. This is because they are actions the central leadership takes with the aim of avoiding future crises or at least reducing their impact.

In 1985 the only non-Han ever to hold the position of CCP secretary in Tibet arrived in Lhasa. This was Wu Jinghua, who was Yi and thus a member of an ethnic minority. Though not Tibetan, he encouraged policies of Tibetanization, renamed streets in Tibetan and attended religious functions himself wearing Tibetan dress.[56]

However, Wu Jinghua was dismissed as CCP TAR secretary, reportedly for "right deviationism," and returned to Beijing in July 1988.[57] His successor was Hu Jintao, who was to become national CCP general secretary in 2002. In December 1992 the central CCP appointed as TAR secretary Chen Kuiyuan, who remained in power until 2000. According to Robert Barnett, Chen should be credited with the "energetic implementation" of the binary policy of development and stability, as well as with raising the standard of living and achieving double-digit economic growth rates while he was in control.[58] He was, however, regarded as a "hard-liner" and demanded absolutely correct ideology in the appointment of cadres. In short, the late 1980s crisis in Tibet led to a CCP leadership in the TAR much less responsive to Tibetan sensitivities than the one in place in the 1980s.

In the early years of the twenty-first century, the post of TAR CCP secretary fell to Guo Jinlong and Yang Chuantang, whom Barnett describes as "relatively mild by local standards."[59] At least they were so by comparison with Zhang Qingli, who took over the position early in 2006. He had been a member of the CCP Central Committee since 2002, and already had experience in that other particularly sensitive ethnic region Xinjiang. Not noted for appreciation of Tibetan culture, it was he who was in power in the TAR during the period of the disturbances in 2008.

There are parallels with the TAR but also slight differences in the pattern of CCP leadership in Xinjiang. The main figure in Xinjiang politics in the last decade of the twentieth century and the first years of the twenty-first was Wang Lequan, the Xinjiang CCP secretary whom we saw earlier as enjoying the reputation for toughness but also intolerance in dealing with ethnic problems, because of an excessive zeal for preventing any hint of separatism. Many blamed him for the 2009 riots, not only because of his inflexibility, but also because of "his failure to adequately anticipate the occurrence of the violence and take proper precautions against it,"[60] which he certainly had the power to do. Following the syringe attacks, an unprecedented Han demonstration in Ürümqi on September 3 demanded his resignation.[61] Wang has good connections in Beijing, which probably helped him stay in office for the time being. Instead, political blame was heaped on the Ürümqi Municipal CCP Secretary Li Zhi, who lost his job not long after the demonstration.

There were some significant developments the following year. In April, Wang Lequan was replaced as Xinjiang CCP secretary by Zhang Chunxian, who had been secretary in Hunan. Politburo Standing Committee member and Vice-President Xi Jinping made a special visit to Ürümqi to announce the decision. Wang received praise from Xi for his loyalty to the CCP and his hard work and was immediately given a senior central CCP post. Meanwhile, his replacement Zhang Chunxian has a very different kind of reputation. In the words of Wu Zhong, China editor for *Asia Times Online*, "outspoken and media-friendly, Zhang was voted by Hong Kong journalists as 'the most open-minded party secretary' on the mainland."[62] In contrast to the situation in Tibet, the central CCP leadership had decided that their principal appointee had become so obsessive over separatism as unnecessarily to irritate local Uygur sensitivities. They clearly hope that a new CCP secretary, even though Han ethnically, may be able to relieve animosities and prevent further crises.

External management of crises in the Tibetan areas and Xinjiang

The principal external method of handling the four crises discussed here was to try and put the blame for them on external forces and win international support for their internal actions. In the case of the Tibetan areas, the external forces focused almost entirely upon the person of the Dalai Lama, who had set up a government-in-exile in Dharamsala, India, after he fled China following the failed 1959 uprising. In Xinjiang, the external forces are much more complex, evolving over the period from 1987 to 2010. Chinese authorities attribute a range of purposes to external forces in stirring up demonstrations and riots, but the main one is to secure independence for Tibet and Xinjiang and by implication the break-up and weakening of China.

The earlier crises

The Chinese government put the main blame for the disturbances of 1987 to 1989 on the Dalai Lama and his Western, especially American, supporters. His winning of the Nobel Peace Prize late in 1989 seemed to the Chinese a direct slap in the face and they reacted furiously. In the 1990s the Dalai Lama made increased numbers of overseas trips, frequently meeting heads of state and government. In July 1996, he gave a major speech in Westminster, advocating for Tibet not full independence, but "genuine autonomy," which would give control over domestic affairs to a Tibetan government but leave foreign affairs to the Chinese. In 1997, a series of motion pictures came out furthering his cause and promoting negative images of Chinese policy and behavior in the Tibetan areas.

The Chinese repeatedly asked foreign leaders not to meet the Dalai Lama, usually to no effect. They also denied accusations of human rights abuses and claimed that what happened inside China was its internal affair, in which foreign powers should not interfere. However, the Tibet lobby focused on the United States became increasingly influential and the Western media overwhelmingly sided with the Dalai Lama and the Tibetan diaspora against the Chinese. Indeed, during the 1990s, he became one of the most significant of all "trend makers," and was able to make Tibetan Buddhism an extremely fashionable religion in many parts of the world.

In Xinjiang, by contrast, there was no figure even remotely comparable to the Dalai Lama who could act as a leader for the Uygur cause. However, there were other circumstances just as important for China. The Baren Township uprising took place only about two months after serious ethnic conflict had erupted in

Dushanbe, capital of Tajikistan, which neighbors Xinjiang. Moreover, although a republic of the Soviet Union, Tajikistan was to be an independent country with the collapse of the Soviet Union at the end of 1991.

The establishment of the Shanghai Five in April 1996 was a cooperative international measure the Chinese leadership could take to oppose the kind of dangers posed in incidents such as the Baren Township uprising.[63] The Ürümqi meeting the preceding month had blamed the disintegration of security in Xinjiang best represented by the uprising on infiltration and subversion "by some hostile forces in the West,"[64] meaning mainly the United States, which many in China still believed was trying to tear China apart. As we saw in the subsection entitled "The structure of crisis management in the Tibetan areas and Xinjiang," the Shanghai Five grew into the Shanghai Cooperation Organization in June 2001, with China very willing to cooperate with the governments of the independent Central Asian states and Russia against terrorism and Islamic fundamentalism in the region. Meanwhile, the September 11 incidents of 2001 added the dimension of the US-led "war on terrorism." It gave China's leaders more opportunity to cooperate with the United States and China against terrorism, and also more reason for clamping down on dissidence in Xinjiang.[65]

An additional factor causing concern to China was the attempt of the faction-ridden Uygur diaspora to recreate itself into a credible opposition able to represent their ethnic group, especially given their lack of counterpart to the Dalai Lama. In Munich in April 2004, the fractious Uygur diasporic associations united into the World Uyghur Congress, which has been able to pull far more weight with Western and other countries than any of its Uygur nationalist predecessors. In November 2006 the World Uyghur Congress elected as its president Rebiya Kadeer, a woman who had formerly been the richest person in Xinjiang but been arrested in 1999 for revealing state secrets to foreigners. Sentenced to eight years in prison, she was released in March 2006 in return for a promise not to involve herself in politics, but went to the United States, where she lives in Washington DC near the White House and has developed as a human rights and democracy activist fiercely opposed to China and its policies and activities in Xinjiang.

The crises of 2008 and 2009

By the time of the crises of 2008 and 2009, both the Tibetan and Uygur causes were thus more internationalized than they had been in 1987,with both diasporas pulling more weight. Meanwhile, China's rise was gathering momentum, with the image of the "China threat" reasserting itself in the West. A major symbol of this rise was the Beijing Olympics of August 2008, for the Chinese a supremely important event into which they invested a great deal of effort and money. In international opinion, the Tibetan disturbances became intertwined both with the Olympic Games and the Olympic Torch which, from March 25 to August 8, was making its way from Athens to Beijing.

The international measures Chinese authorities took to manage the crisis posed by the Tibetan riots and disturbances of 2008 fall into several categories.

The first and most obvious was that the Chinese leaders did all they could to divert the blame for the riots away from themselves and towards the "Dalai clique." This phrase meant mainly the Dalai Lama himself, who immediately responded that he did not support any violent riots, but sympathized with the Tibetans, because the root cause of the disturbances was Chinese oppression. The "Dalai clique" also included a variety of other Tibetan diasporic groups the Chinese leadership and media associated with the Dalai Lama, foremost among them the Tibetan Youth Congress, which advocates full independence from China, not merely the "genuine autonomy" espoused by the Dalai Lama.

Second, the Chinese tried to counter hostile comment from overseas that resulted both from the disturbances in the Tibetan areas and threats against the Olympic Torch. The Chinese state assisted with money and organization a worldwide, largely spontaneous, nationalist Chinese upsurge, which aimed to protect the Olympic Torch, especially after an attack on it in Paris. Chinese representatives criticized several world leaders who announced that they would boycott the Olympic Games Opening Ceremony unless the Chinese held talks with the Dalai Lama as a step towards solving the Tibet problem. Foremost among them was French President Nicolas Sarkozy.

A nationalist wave spread across China. Temporary boycotts affected organizations believed to be supporting the Dalai Lama, the most prominent example being against the French supermarket chain Carrefour. Though they have not been able to persuade Western opinion to support their view on Tibet, the Chinese authorities have been much more successful in convincing the great majority of their own people. Warren Smith is quite right to comment:

> The Tibetan optimism that the Chinese people, if not their government, are becoming more sympathetic to the Tibetan issue is also contradicted by the rise of nationalism among Chinese of all types, whether Buddhists or democracy advocates or those who have travelled to or worked in Tibet.[66]

In April 2008, the Chinese did indeed offer talks with the Dalai Lama's representatives, as Sarkozy and others had demanded, and it was probably part of their external management of the crisis in the Tibetan areas. Although they held to their view that nobody should interfere in China's internal affairs and denied that their offer was related to pressures from people like Sarkozy, they probably felt they had nothing to lose by trying to assuage public opinion outside China. Later in the year, the Dalai Lama's representatives handed a memorandum to the Chinese with a detailed proposal concerning the future of Tibet. Unfortunately, it was very little different from what they had been suggesting for some time, and in some ways even increased their demands. In any case, the Chinese scoffed at it, issuing a formal rejection in November. By that time the Olympic Games were over, having apparently enjoyed the success the organizers had wanted. Sarkozy had attended the Opening Ceremony, along with numerous other foreign leaders.

As a way of managing any crisis involving Tibet, pressure against the Dalai Lama is ongoing. Many foreign leaders have exercised more caution than before

2008, albeit reluctantly. Moreover, China has succeeded in gaining concessions from some countries in Tibet policy. When Sarkozy met the Dalai Lama in Poland early in December 2008, against China's express wishes, China retaliated by cancelling the EU–China summit slated to take place in Lyon, France, shortly after. It was as a result that, following consultations between the foreign ministries of the two countries, France and China issued a communiqué in April 2009, in which France made a point of recognizing Tibet as "an integral part of the Chinese territory," a policy it said it would never change. In the communiqué France conceded that it "refuses to support any form of 'Tibet independence.'"[67]

As in Tibet, one way the Chinese government managed the 2009 crisis in Xinjiang was to blame outside forces. These included al Qaeda and the United States Congress.[68] However, this time the chief target was Rebiya Kadeer, and the World Uyghur Congress of which she is president. Chinese media and authorities described her as a terrorist and a criminal, likening her to the Dalai Lama, and mounted an extensive international campaign to discredit her. However, the Chinese attempt proved counter-productive: it gave her a good reputation she lacked before, and while hardly anybody knew of her before the riots quite a few learned her name because of the Chinese propaganda campaign against her. She is still a very insubstantial figure by comparison with the Dalai Lama and, unlike him, has no pretension to being a religious leader. It is worth noting that the documents from the May 2010 work conference on Xinjiang do not even mention her.

Evaluation and conclusion

We can now apply criteria from China's imperatives to reach an evaluation of China's crisis management in the relevant ethnic minority areas and to speculate briefly about the future.

Since China's top imperative is to remain united, the first question is whether separatism has become stronger or weaker as a result of China's handling of the crises under discussion in this chapter. The answer is probably that it is weaker. Although there are still forces, both inside and outside China, that seek to undermine China's unity, the handling of the crises suggests that China's determination and ability to resist them has grown more strongly over the decades from the late 1980s to 2010 than any separatist trends. China cannot control the activities of external enemies whom they perceive as threatening its unity, but its international strategic and economic impact is on the rise and probably enough to overcome counter influences. It follows that China's crisis management has lessened the potential for national collapse, not increased it.

If we consider China's imperative of stability and the "harmonious society," we can only reach a different conclusion, namely that China has become more unstable and less harmonious in the relevant areas than it used to be. The disturbances and riots suggest an unstable and quite disharmonious society. A judgment on the extent of disharmony will vary depending on whether one blames internal or external causes, but either way, China's short-term crisis management clearly failed to progress the Tibetan and Uygur areas towards a harmonious society. The ethnic 2008 Tibetan disturbances and the rioting in Ürümqi can only cause more bitterness, which China's initial response probably exacerbated.

On the other hand, the methods of crisis management and the resultant tightening of security could reduce the possibility of further disturbances. Chinese security forces feel they were caught by surprise in 2008 and 2009 and are determined that will not happen again. Moreover, the jury is still out on whether the longer-term measures adopted at the January and May 2010 meetings on the Tibetan areas and Xinjiang can ease the situation. It may be possible to find ways of giving more attention to ethnic sensitivities and trying to cooperate with ethnic cultures, rather than appearing to regard them as inimical to Chinese progress. There is virtually no possibility of major short-term improvement in ethnic relations, but it cannot be ruled out over the next decade or so.

On economic progress, the record seems fairly positive. Apart from adopting policies prevalent throughout the country, such as the campaign to "open up the west," China's crisis management has included several fairly comprehensive measures to promote economic growth in places that include the Tibetan areas and Xinjiang, The measures adopted at the 2010 meetings have the potential to increase growth further. On the other hand, inequality has been a serious problem and could become even more so unless effective counter-measures are adopted. The government has to reject the alternative of simply allowing the Tibetan areas and Xinjiang to remain poor, while the rest of the country develops. But it may be possible to find ways of involving the Tibetans and Uygurs more in this development, making them feel that they are genuinely part of China's successes.

It appears to follow from the record that China's crisis management in the Tibetan areas and Xinjiang has been mixed but dominantly effective. Economic growth should continue there and the break-up of China is unlikely in the foreseeable future. If a state collapse or party overthrow should occur, it will not be due to the failure of crisis management in the Tibetan areas and Xinjiang but to other factors.

Notes

1 PRC National Bureau of Statistics (comp.), *Zhongguo tongji nianjian 2006* [*China Statistical Yearbook 2006*] (Beijing: China Statistics Press, 2006), p. 102.
2 "Population hits 1.30756 bln by 2005: Survey," *People's Daily Online*, March 17, 2006, at http://english.people.com.cn/200603/17/eng20060317_251506.html (Retrieved November 8, 2009).
3 "Constitution of the People's Republic of China," *Beijing Review*, Vol. 25, No. 52 (December 27, 1982), pp. 11, 12.
4 See *Zhongguo renmin gongheguo minzu quyu zizhi fa* [*Law of the People's Republic of China on Regional National Autonomy*] (Beijing: Minzu chubanshe, 2001), pp. 2 and 4 (Chinese) and 28 and 32 (English).
5 See also the discussion in Christian Tyler, *Wild West China: The Taming of Xinjiang* (New Bruswick, NJ: Rutgers University Press, 2004), pp. 206–207.
6 Graham Young and Yingjie Guo, "Managing Rights Talk in the 'Harmonious Society'," in Joseph Y.S. Cheng (ed.), *Challenges and Policy Programmes of China's*

New Leadership (Hong Kong: City University of Hong Kong Press, 2007), p. 97. See also David Shambaugh, *China's Communist Party, Atrophy and Adaptation* (Berkeley and Los Angeles: University of California Press, 2008), pp. 115–118.

7 Zhu Rongji, "Report on the Work of the Government," *Beijing Review*, Vol. 43, No. 14 (April 3, 2000), p. 21.

8 David Goodman, "The Campaign to 'Open Up the West': National, Provincial-level and Local Perspectives," in David Goodman (ed.), *China's Campaign to "Open Up the West": National, Provincial and Local Perspectives* (The China Quarterly Special Issues, New Series, No. 5, Cambridge: Cambridge University Press, 2004), p. 12.

9 There is now an extensive literature on PRC policy towards its ethnic minorities. For the period until the early 1990s see, for example, Colin Mackerras, *China's Minorities, Integration and Modernization in the Twentieth Century* (Hong Kong: Oxford University Press, 1994), pp. 139-166 and for the period from 1990 to 2002 see Colin Mackerras, *China's Ethnic Minorities and Globalisation* (London and New York: RoutledgeCurzon, 2003), pp. 37–55.

10 See Article 17 of the autonomy law as amended in 2001 in *Minzu quyu zizhi fa*, pp. 7–8 (Chinese) and 39 (English).

11 See Article 49 of the amended law in *Minzu quyu zizhi fa*, pp. 16 (Chinese) and 55 (English).

12 See "Constitution of the People's Republic of China," p. 16.

13 See the territory explained and shown in two maps in the official website of the Central Tibetan Administration, at http://www.tibet.net/en/index.php?id=13&rmenuid=8 (Retrieved November 8, 2009).

14 The figures come from or are based on Department of Population, Social, Science and Technology Statistics, National Bureau of Statistics, PRC & Department of Economic and Development, State Ethnic Affairs Commission, PRC (comp.), *2000nian renkou pucha Zhongguo minzu renkou ziliao* [*Tabulation on Nationalities of 2000 Population Census of China*], 2 Vols (Beijing: Minzu chubanshe, 2003), especially Vol. 1, pp. 2–6.

15 The figures are calculated from data in Population Statistics Office of the State Statistical Bureau (comp.), *Zhongguo renkou tongji nianjian 1990* [*China Population Statistics Yearbook 1990*] (Beijing: Kexue jishu wenxian chubanshe, 1991), especially pp. 78–79.

16 Robert F. Ash (comp.), "Quarterly Chronicle and Documentation (April–June 1990)," *The China Quarterly*, No. 123 (September 1990), p. 583.

17 Tyler, *Wild West China*, p. 165.

18 An official CCP view, which is very clear that the rebellion was an attempted *jihad*, is contained in a report compiled by the Research Group Council of the CCP. Entitled "The Anti-Separatism Struggle and Its Historical lessons Since the Liberation of Xinjiang" and dated August 1993; the report is cited by Tyler in *Wild West China*, pp. 165–166. For an account by a strong supporter of the uprising, and also claiming it as *jihad*, see Michael Winchester, "Beijing vs. Islam," *Asiaweek*, Vol. 23, No. 42 (October 24, 1997), p. 31. There is a growing literature on the Baren Township uprising and later disturbances. For instance, see James A. Millward, *Eurasian Crossroads: A History of Xinjiang* (New York: Columbia University Press, 2007), pp. 325–328.

19 Robert Barnett, "Thunder from Tibet," *The New York Review of Books*, Vol. 55, No. 9 (May 29, 2008), web version at http://www.nybooks.com/articles/21391 (Retrieved January 23, 2009).

20 Robert Barnett, "The Tibet Protests of Spring, 2008: Conflict between the Nation and the State," *China Perspectives*, No. 3 (2009), pp. 8–9.

21 "Questions, Answers about Casualties, Damages of Recent Riots," *China View*, March 25, 2008, website at http://news.xinhuanet.com/english/2008-03/25/content_7857168.htm (Retrieved December 4, 2009).

22 See "Press Statement," April 29, 2008, in the official website of the Central Tibetan Administration, at http://www.tibet.net/en/index.php?id=33&articletype=press&rmenuid=morepress&tab=2#TabbedPanels1 (Retrieved December 5, 2009).

23 Xinhua, "Innocent Civilians Make up 156 in Urumqi Riot Death Toll," *China View*, August 5, 2009, at http://news.xinhuanet.com/english/2009-08/05/content_11831350.htm (Retrieved November 29, 2009).
24 Kristine Kwok, "Thousands Protest over Urumqi Syringe Attacks," *South China Morning Post*, September 4, 2009.
25 Michael Wines, "A Strongman Is China's Rock in Ethnic Strife," *New York Times*, July 10, 2009, at http://nytimes.com/2009/07/11/world/asia/11xinjiang.html?_r=1 (Retrieved July 6, 2010).
26 Richard M. Bennett, "Tibet, the 'Great Game' and the CIA," *Asia Times Online*, March 26, 2008, at http://www.atimes.com/atimes/China/JC26Ad02.html (Retrieved July 6, 2010). Bennett is intelligence and security consultant, AFI Research.
27 Amalendu Misra, "Shanghai 5 and the Emerging Alliance in Central Asia: The Closed Society and Its Enemies," *Central Asian Survey*, Vol. 20, No. 3 (September 2001), p. 305.
28 Tai Ming Cheung, "Guarding China's Domestic Front Line: The People's Armed Police and China's Stability," *The China Quarterly*, No. 146 (June 1996), p. 527.
29 Ibid., p. 535.
30 Ibid., p. 538.
31 See Ronald D. Schwartz, "The Anti-Splittist Campaign and Tibetan Political Consciousness," in Robert Barnett (ed.), *Resistance and Reform in Tibet* (London: Hurst, 1994), p. 218.
32 A. Tom Grunfeld, *The Making of Modern Tibet*, Rev. ed. (New York: M.E. Sharpe, 1996), p. 233.
33 Schwartz, "The Anti-Splittist Campaign and Tibetan Political Consciousness," p. 211.
34 Ibid., p. 207.
35 Winchester, "Beijing vs. Islam," pp. 31, 34.
36 That the Chinese authorities believed that this kind of ideology motivated the uprising is confirmed in the official report cited in Tyler, *Wild West China*, pp. 165–166.
37 Xinhua, "Innocent Civilians Make up 156 in Urumqi Riot Death Toll."
38 Human Rights Watch, *"We Are Afraid to Even Look for Them": Enforced Disappearances in the Wake of Xinjiang's Protests* (New York: Human Rights Watch, 2009).
39 For instance, see Josephine Ma, "Weeping Monks' Cry for Freedom Gives the Lie to Peace in Lhasa," *South China Morning Post*, March 28, 2008, at http://archive.scmp.com/showarticles.php (Retrieved December 5, 2009).
40 Graham E. Fuller and Jonathan N. Lipman, "Islam in Xinjiang," in S. Frederick Starr (ed.), *Xinjiang: China's Muslim Borderland* (New York and London: M.E. Sharpe, 2004), p. 325.
41 Melvyn C. Goldstein, *The Snow Lion and the Dragon: China, Tibet, and the Dalai Lama* (Berkeley: University of California Press, 1997), pp. 80–83.
42 Zhao Shenying (ed.), *Zhonggong Xizang dangshi da shiji, 1949–1994* [*Chronology of Party History in Tibet under the CCP, 1949–1994*] (Lhasa: Xizang Renmin Chubanshe, 1995), pp. 456–459.
43 This material is based on an internal report: Yang Faren, Li Ze and Dong Sheng (eds), *Fan Yisilanzhuyi, fan Tujuezhuyi yanjiu* [*Research on Pan-Islamism and Pan-Turkism*] (Ürümqi: Xinjiang Shehui Kexue Yuan, 1994). The quotation is taken from Nicolas Becquelin, "Xinjiang in the Nineties," *The China Journal*, No. 44 (July 2000), p. 69 and uses his translation.
44 BBC Monitoring Service: Asia Pacific, May 9, 1996, from Xinjiang Television, Ürümqi, in Chinese, evening report of May 6, 1996.
45 Winchester, "Beijing vs. Islam," pp. 36, 40.
46 HRIC (Human Rights in China), "Devastating Blows, Religious Repression of Uighurs in Xinjiang," *Human Rights Watch*, Vol. 17, No. 2 (C) (April 2005), p. 58.

47 See *Xinjiang Daily*, May 7, 1996, as quoted by Jane Macartney, ''China Moslem Region Lashes out Against Separatism,'' *Reuter News Service*, May 12, 1996. *Xinjiang Daily* of May 20, 1996 also suggested introducing more Han cadres to complement minority counterparts. See Robert F. Ash (comp.), ''Quarterly Chronicle and Documentation (April –June 1996),'' *The China Quarterly*, No. 147 (September 1996), p. 1026. It may be added that the training of ethnic cadres, and especially those from Xinjiang and Tibet are part of the cadre programme at the Central Party School. According to Frank N. Pieke, *The Good Communist, Elite Training and State Building in Today's China* (Cambridge: Cambridge University Press, 2009), p. 67, there are courses lasting two years especially for Tibetan and Xinjiang cadres, which are ''aimed at students who have not previously studied at university'' with an undergraduate degree ''given at the end of the course.''

48 Minnie Chan and Kristine Kwok, ''After Rioting, PLA Steps Up Efforts to Recruit Minorities; Army Targets Candidates with Knowledge of Tibet, Xinjiang,'' *South China Morning Post*, December 11, 2009, p. 4.

49 Chris Buckley (editing by Nick Macfie and Valerie Lee, for *Reuters*), ''China Launches Education Drive in Lhasa, Warns Citizens,'' *AlertNet: Alerting Humanitarians to Emergencies*, April 21, 2008, at http://www.alertnet.org/thenews/newsdesk/PEK289977.htm (Retrieved December 5, 2009).

50 Information Office of the State Council, ''China's Ethnic Policy and Common Prosperity and Development of All Ethnic Groups,'' *China View*, September 27, 2009, at http://news.xinhuanet.com/english/2009-09/27/content_12117708_2.htm, p. 3 (Retrieved July 8, 2010).

51 Shi Jiangtao, ''After Unrest, Beijing Vows Tibet Spending, Plan Includes Tibetan Areas in Other Provinces,'' *South China Morning Post*, January 23, 2010, web version at http://www.scmp.com/portal/site/SCMP/menuitem.2af62ecb329d3d7733492d9253a0a0a0/?vgnextoid=429cc29090756210VgnVCM100000360a0a0aRCRD&ss=China&s=News (Retrieved January 26, 2010).

52 Willy Lam, ''Hu's 'New Deal' with Tibet: Chinese Characteristics and Tibetan Traits?'' *China Brief*, Vol. 10, No. 2 (January 21, 2010), web version at http://www.jamestown.org/programs/chinabrief/single/?tx_ttnews[tt_news]=35930&tx_ttnews[backPid]=25&cHash=32e2b6280c (Retrieved January 26, 2010).

53 See Xinhua report of January 23, 2010, quoted in ''Communist Party Leadership Outlines 2010-2020 'Tibet Work' Priorities at 'Fifth Forum,''' a synthesis of the Symposium dated March 9, 2010 put out by Congressional – Executive Committee on China. See http://www.cecc.gov/pages/virtualAcad/index.phpd?showsingle=135152 (Retrieved July 8, 2010).

54 Cui Jia and Zhu Zhe, ''Xinjiang Support Package Unveiled,'' *China Daily*, May 21, 2010, p. 1. The quotations are the journalists' paraphrase of Hu's words.

55 Ibid., p. 2. The journalists are here quoting Ashar Turson's words directly.

56 Grunfeld, *The Making of Modern Tibet*, p. 225.

57 See ''Appendix. Provincial-Level Leaders in Tibetan Areas Since 1950,'' in Barnett (ed.), *Resistance and Reform in Tibet*, p. 296.

58 Robert Barnett, ''Chen Kuiyuan and the Marketisation of Policy,'' in Alex McKay (ed.), *Tibet and Her Neighbours: A History* (London: Thames & Hudson Edition Hansjörg Mayer, 2003), p. 231.

59 Barnett, ''The Tibet Protests of Spring, 2008,'' p. 22.

60 Shi-ren Hou, '''King of Xinjiang' Faces Blame for Riots,'' *Asia Times Online*, July 16, 2009, at www.atimes.com (Retrieved November 29, 2009).

61 Kwok, ''Thousands Protest over Urumqi Syringe Attacks,'' p. 1.

62 Wu Zhong, ''Greater China: Beijing Seeks a Fresh Start in Xinjiang,'' *Asia Times Online*, May 19, 2010, http://www.atimes.com/atimes/China/LE19Ad01.html (Retrieved May 23, 2010).

63 See also Pan Guang, "China in the Shanghai Cooperation Organization," in Wang Gungwu and Zheng Yongnian (eds), *China and the New International Order* (London and New York: Routledge, 2008), pp. 237–255.

64 See Macartney, "China Moslem Region Lashes out Against Separatism."

65 There is now a good and growing literature on these topics, which are not focal to this chapter. For instance, see Nicolas Becquelin, "Staged Development in Xinjiang," in Goodman (ed.), *China's Campaign to "Open Up the West,"* pp. 44–64; Ann McMillan, "Xinjiang and Central Asia, Interdependency – Not Integration," in Colin Mackerras and Michael Clarke (eds), *China, Xinjiang and Central Asia: History, Transition and Crossborder Interaction into the 21st Century* (London and New York: Routledge, 2009), pp. 94–114; and Fuller and Lipman, "Islam in Xinjiang," pp. 338–352.

66 Warren W. Smith Jr., *China's Tibet? Autonomy or Assimilation* (Lanham: Rowman & Littlefield, 2008), p. 287.

67 "Full Text of Press Communique between China and France," *China Daily*, April 1, 2009, web version at http://www.chinadaily.com.cn/china/2009-04/01/content_7640216.htm.

68 Human Rights Watch, *We Are Afraid to Even Look for Them*, p. 13.

5 Managing pandemic/epidemic crises

Institutional setup and overhaul

Hongyi Lai

In April 2003 the continued rampage of the severe acute respiratory syndrome (SARS) triggered the most systemic reform of the management of epidemic and other crises in China. In April 2003 the Chinese leadership launched an open campaign to fight against SARS.[1] In the subsequent years China's institutions for managing major crises (including but beyond epidemics) have been overhauled. China's change in the management of pandemic and epidemic crises thus provides a critical glimpse of the major overhaul in its crisis management in the past decade.

April 2003 thus becomes a watershed in China's epidemic management. This chapter examines China's epidemic management before and after April 2003. For both periods, the following elements integral to epidemic management will be examined: laws and regulations which make up the legal framework; classification of epidemics; and the types of agencies that were entrusted with the duties. Furthermore, for both periods the three stages of epidemic management will be reviewed, i.e., preventive, monitoring, and containment stages. In particular, this chapter will analyze epidemic management institutions. Institutions refer to the legal and administrative framework, entrusted jurisdiction and duties of agencies and units, and prescribed and practiced norms at preventive, monitoring, and containment stages.

The contrast in epidemic management, especially its institutions before and after April 2003 can be briefly highlighted here. Prior to April 2003 the Party leadership emphasized economic growth and social stability. As a result, the administration and the health department (the Ministry or Bureau of Health) managed the epidemics secretively and were reluctant to report epidemics publicly and respond swiftly and firmly to them. Adding to this defect were poor interagency coordination and the loopholes in the 1989 epidemic law and official accountability. Since April 2003 these institutional deficiencies have been incrementally addressed and national and local epidemic crisis management institutions have been strengthened. As a result, epidemic management has become more open, swift, and effective.

Pandemic/epidemic management prior to April 2003

The following analysis will start with the relevant laws and regulations, classification of epidemics, and agencies for managing epidemics. In a sense these three components help to define the norms for epidemic management. The analysis will then move on to cover the preventive, monitoring, and containment stages. In the analyses both the prescribed norms and actual practice will be examined so as to provide a complete picture of epidemic management in China.

Legal framework of epidemic management

The epidemic management institutions prior to April 2003 were formally stipulated by the Epidemic Law of 1989 and the Implementing Measures of 1991. The Law on Prevention and Control of Epidemics (the Epidemic Law in short) came into effect in February 1989. The Ministry of Health promulgated the Implementing Measures for the Law on Prevention and Control of Epidemics (the Implementing Measures in short) in December 1991. The Epidemic Law of 1989 was the main law of the People's Republic of China regarding epidemic management, while the Implementing Measures was the main policy document regarding epidemic management.[2] Both documents, especially the Epidemic Law, stipulated epidemic management institutions. The Epidemic Law spelled out the duties and measures for each level of the government regarding epidemic management, including early warning and monitoring, reports, and management.[3]

Classification of epidemics

In particular, the Law stipulated three categories of epidemics. Class A included plague and cholera. Class B included 22 epidemics, such as AIDS, anthrax, dengue fever, typhoid and paratyphoid, gonorrhea, syphilis, measles, diphtheria, scarlet fever, epidemic hemorrhagic fever, rabies, and malaria. Class C included 11 epidemics, including pulmonary tuberculosis, schistosomiasis, filariasis, echinococcosis, leprosy, and influenza (Article 3). The Law authorized the State Council to add or remove Class A epidemics and the Ministry of Health to add or remove Class B and Class C epidemics (Article 3).

Agencies for managing epidemics

The prescribed agencies for epidemic managements in the Epidemic Law of 1989 and the Implementing Measures of 1991 can be summarized as follows. Under the leadership of the Party Committee, especially the Party Secretary, the administrative branch at all levels was responsible for managing epidemics and announcing major anti-epidemic measures under their jurisdiction, and the health department was in charge of coordination of epidemic management by health bureaus and medical units. Below these governmental bodies, disease-control agencies played a key professional role in monitoring and detecting epidemics, and local medical units or disease control units took charge of quarantining patients with class-A or other epidemics, treating patients and suspected patients, and properly handling epidemic waste and items.

1. Overarching epidemic management agencies

The administrative branch at all levels, i.e., the State Council at the national level and the People's Government at the local level, was responsible for the overall epidemic management in the areas under its jurisdiction. Its role included inter-departmental coordination, drafting of epidemic plans, and organization of the implementation of the plans. Local administration could also adopt emergency measures for epidemic management after reporting them to the next higher level, whereas the State Council could adopt emergency measures directly (see Articles 4, 5, 9, 25, and 26, the Epidemic Law of 1989). In practice, the epidemic management was subject to the influences and instruction of the Party (especially the Party Committee and the Party Secretary) at the same level of the state. This was in line with the principle of the Party leadership. The Party or the Party Secretary could issue orders or guidelines on how to manage an epidemic.

In addition, in practice small groups might be established to coordinate epidemic management, provide technical guidance, and offer advice on etiological tests. The SARS management in Guangdong served as an example. In early February 2003 the Guangdong authority established the Guangdong Provincial Coordinating Group for Preventing and Controlling Unexplained Pneumonia, a Provincial Small Group of Experts for Guidance concerning Medical Aid, a Small Group for Guidance in Investigation in Epidemiology, and a Small Group for Technical Guidance at Etiological Tests. Moreover, the authority urged all localities to form corresponding groups and report their memberships and contact information by February 20.

2. Top medical agencies in charge of epidemic management

The health department took care of medical aspects of epidemic management. They included supervision and inspection of prevention, control, monitoring, and containment of epidemics; issuing of an order to any units or individuals to improve their anti-epidemic efforts; and imposition of compulsory measures on units or individuals who failed to adopt necessary anti-epidemic sanitary measures. The health department was responsible for reporting the epidemic conditions to the health department at the next higher level and the administration at the same level (Articles 5, 23, 27, 32, and 35, the Epidemic Law of 1989).

Meanwhile, the animal husbandry and veterinary departments of governments at each level were entrusted with responsibility for preventing, treating, and controlling infectious diseases of domestic animals and fowl that could be transmitted to human beings and animals (Article 18, the Epidemic Law of 1989).

3. Medical agencies providing technical expertise and guidance for epidemic management

The disease-control agencies, which later became the Centers for Disease Control and Prevention (CDCs), monitored, tested, and tracked the epidemic conditions, reported epidemic conditions to the health departments, and guided and supervised the anti-epidemic work. They might also participate in the quarantine of infectious patients (Articles 5, 15, 17, 24, and 27, Epidemic Law of 1989).

4. Medical units directly involved in epidemic management

Hospitals and clinics were directly involved in the prevention and control of epidemics and treatment and quarantine of patients. They should report epidemic cases to the local CDC. Medical suppliers were supposed to furnish sufficient equipment and medicine (Articles 5, 11, 17, 24, 27, and 35, Epidemic Law of 1989).

Prevention of epidemics

In China efforts for preventing epidemics have always involved usually the government, the medical and health institutions, other social units and individuals. These efforts have included but are not limited to immunization programs, educational campaigns, sanitary measures, and proper procedures at medical units for minimizing virus transmission. These efforts will be detailed concisely below.

For decades the Chinese government has implemented immunization among children. Building on its immunization efforts from the 1950s to 1978, the Chinese government furthered its children immunization program in the reform era. By 1988, as much as 85 percent of children in all provinces were immunized against tuberculosis, poliomyelitis, measles, whooping cough, diphtheria, and tetanus. In 1990 and 1995, the majority of children in each county and township were immunized against the aforementioned diseases. In addition, efforts were made to contain the spread of hepatitis B, and by 1997, 50 percent of the children in China under one year old had been injected with the hepatitis B vaccine. Mass immunization of the population against these diseases (most of which were on the aforementioned list of Class B and Class C epidemics) helped to reduce the incidents of these infectious disease.[4]

In addition, in the reform era the state continued the sanitary campaigns that it introduced in the 1950s. During the campaign work units, neighborhood committees, and governmental organizations in all localities, especially cities, would be mobilized to clean up streets and buildings and kill four pests, namely, rodents, flies, mosquitoes, and cockroaches, as they were believed to transmit infectious diseases. In addition, localities were asked to establish or maintain public health facilities, provide better management of sewage, waste, and feces, and improve the hygienic condition of drinking water so that it met the national standard. In addition, in 1980 China joined the United Nations project for clean water by introducing tap water to the countryside. By 1995 tap water was said to reach 43.5 percent of the rural population. These efforts have helped to improve the sanitary conditions of the population and contain epidemics.[5] All municipalities, municipal districts, and counties were required to designate hospitals, clinics, and wards for treating infectious diseases (Article 11, Epidemic Law of 1989).

However, during the SARS campaign it was revealed that the above requirement was not met, as many cities and counties lacked a designated hospital or ward to handle epidemic cases.

In addition, health departments at each level were required to have health organizations or personnel to take charge of prevention and control of infectious diseases for their respective units or their respective communities (Article 11, Epidemic Law). Three health stipulations were also imposed by the national government through the Epidemic Law to minimize the spread of epidemics. First, infectious disease patients, pathogen carriers, and suspected infectious disease patients were barred from jobs which the health administration department suspected could cause the spread of epidemics (Article 14). In addition, all health-related institutions or units with samples of pathogenic microorganisms were required to rigorously abide by the requirements and operation procedures stipulated by the health department under the State Council (Article 15) in order to prevent release of the pathogen. Third, units should take effective measures to protect medical professionals engaged in anti-epidemic work from being infected (Article 20). However, again, as seen in SARS when an epidemic was largely new these cautious preventive measures might not be well implemented. Sometimes medical professionals might let their guard down as they were unfamiliar with the disease.

Monitoring of epidemics

According to the 1989 Epidemic Law the health department (the Ministry of Health at the national level and the Health Bureau at the local levels), as well as disease-control agencies, took charge of monitoring and inspecting epidemics, and supervising the medical and health work related to epidemics. Article 21 of the 1989 Epidemic Law also required immediate reporting of epidemic conditions by informed individuals to nearby medical care institution, and health department, disease-control agencies. The article also required medical, health, or anti-epidemic personnel on duty to report the following epidemic situations to the local health and anti-epidemic agency within the time limit prescribed by the health administration department under the State Council: patients, pathogen carriers, or suspected patients of Class A or Class B infectious diseases, or patients, pathogen carriers, or suspected patients of Class C infectious diseases. When a health and anti-epidemic agency found the spread of infectious diseases or received a report on the epidemic occurrence of Class A epidemic or of AIDS or pulmonary anthrax (Class B infectious disease), it should immediately report to the local health department, which should immediately report to the local government and the health department under the State Council. Article 22 of the Epidemic Law also prohibited governmental officials or health personnel from covering up or making a false report on the epidemic situation. Article 23 required the national health department to promptly release and publicize the epidemic situation and authorized local health departments to publicize local epidemic conditions.

In December 1991 the Ministry of Health issued the Implementing Measures for the Law on Prevention and Control of Epidemics. It imposed time limits for epidemic reporting and categorized AIDS and smallpox under Class A according to the nature of germs and viruses.[6]

During the SARS epidemic from January to early April, however, localities and the military were slow to report the SARS conditions to the national government. Some of them even noticeably underreported the SARS statistics.

Guangdong during the early months of the outbreak of SARS provides an example regarding the monitoring of epidemics. Since mid December 2002 a mysterious respiratory disease (which later was confirmed to be SARS) took place in Heyuan City. The city authority reported it to the Guangdong Bureau of Health (GDBH) only in early January 2003. After mid January 2003 the GDBH received a report on a mysterious infectious disease from Zhongshan City. In both cases the GDBH and Center for Disease Control and Prevention of Guangdong Province (GDCDC) dispatched a team of experts to investigate. The experts coined the disease atypical pneumonia (ATP) (or unexplained pneumonia).[7] Days before and during the Chinese New Year in late January the SARS cases grew rapidly in Guangdong. On February 7 the Acting Vice Governor Li Hongzhong submitted the SARS report from the GDBH to the Party Committee and the People's Government of Guangdong Province.

Containment of epidemics

The 1989 Epidemic Law authorized local government to take the following dramatic measures to contain epidemics after reporting to a higher-level of government – suspension of markets, assemblies, and cinema shows; closure of schools and factories and halt of business operations; temporary requisition of houses and transport vehicles; closure of contaminated drinking sources (Article 25). It also authorized the local government at the county level or above to impose health inspections on people or goods entering or leaving epidemic areas after reporting to the government at a higher level. Furthermore, it authorized provincial government to impose quarantine on areas of Class A epidemics within the province and authorized the State Council to impose quarantine on epidemic areas across provinces or quarantine the national border (Article 26). But in reality the local government was reluctant to take measures that appeared to be very dramatic, such as suspension of markets and business and closure of schools and factories, as it would scare away tourists and investors and dampen local economic activities. This was seen in the SARS management in China prior to April 2003.

Article 24 of the Epidemic Law also required medical, health, and disease control personnel to quarantine patients and pathogen carriers of Class A infectious diseases and patients of AIDS and of pulmonary anthrax (Class B infectious disease) for treatment; take necessary treatment and control measures for patients of Class B and Class C epidemics (other than AIDS and pulmonary anthrax); and keep suspected patients of any Class A infectious disease under medical observation in designated places until the results of a diagnosis were available. These personnel should take necessary sanitary disposal and preventive measures regarding places and objects contaminated by patients, pathogen carriers, and suspected patients of infectious diseases and persons in close contact with them. Concerned patients of infectious diseases, and their relatives, friends, and work units, as well as the local organizations, were required to cooperate.

Again, the SARS management in Guangdong offers an illustrative example regarding the responses to epidemics by local governments and local health departments. On January 23, 2003 the GDBH issued a document to all health agencies at the prefecture level and above. It notified them that SARS (coined ATP) had occurred in the province, and advised these agencies to take preventive measures and quarantine patients. The document, however, was delayed for days and could only be viewed by senior officials at each recipient agency.[8] On February 3 the GDBH issued a document, calling on all hospitals to report SARS cases to CDCs and all CDCs to report SARS cases to the CDC of the next higher level.[9]

Before and during the Chinese New Year the SARS cases grew rapidly in Guangdong. On February 10, amidst widespread rumors, panic purchasing took place.[10] On February 11, Zhang Dejiang, Guangdong Party Secretary, inspected Guangdong Bureau of Health. Zhang set the following three guidelines: no efforts should be spared to save patients; the cause of the epidemic should be quickly ascertained and the epidemic should be effectively contained; normal work, life, and social order should not be disturbed.[11] Upon Zhang's suggestion, Guangzhou and Guangdong Bureaus of Health (GZBH and GDBH) hosted a live-televised press conference on the same day, and maintained that the epidemic was well under control.[12]

However, on February 12, panic purchasing occurred again in a dozen of the major cities in Guangdong. In response to the criticisms by national leaders, the administration and the Party committee in Guangzhou and Guangdong published SARS statistics and allowed temporary open discussion in media. Governmental departments in charge of prices, industry, and commerce mobilized a large quantity of goods to stabilize supplies and prices. Governmental agencies organized daily disinfection and cleaning of public places such as shopping malls, railway stations, taxi stations, and bus stations. However, no quarantine and health checks at the border or mass transport hubs were imposed. Many travelers to and out of Guangdong were not aware of potential risks of infections, which resulted in the transmission of SARS outside Guangdong to Shanxi, Beijing, Hong Kong, Southeast Asia, and North America.[13] Moreover, open discussion lasted about two weeks. During February 11–16, the Department of Propaganda (DOP) the Party Committee of Guangdong mandated the news outlets throughout the province to follow its scripts in reporting SARS. In addition, from then on statistics on SARS were kept secret.[14]

Thus while the authority in China acted to cope with epidemics, it was unwilling to push for higher-profile measures and publicize the epidemic conditions and anti-epidemic methods. This severely limited the effectiveness of anti-epidemic efforts.

Drawbacks of epidemic management prior to April 2003

It is clear from the aforementioned discussion and management of SARS in the early months that several noticeable deficiencies existed in the epidemic management institutions prior to April 2003.[15] They related to the reluctance of the authority in charge in undertaking dramatic and decisive measures to contain epidemics and in publishing truthful epidemic statistics. They also concerned coordination of agencies and units for epidemic management across regions and soundness of laws and regulations.

1. Political concerns

Management of epidemics could be undermined by political concerns, especially the need to showcase stability. Prior to April 2003 social stability and economic growth were paramount concerns for local Party and even administrative leaders. These concerns led to a downplay of epidemic conditions and underestimation of destruction from epidemics. This defect was shown clearly in the initial management of the SARS.[16]

2. Restriction of Information

The usual approach of governmental and even medical agencies was to restrict information on the epidemic conditions to senior medical officials at all levels and downplay the epidemic conditions to the public, despite that it violated the Epidemic Law (Articles 22 and 23). These agencies did so out of fear of affecting economic growth or derailing major political events. As late as early April the Ministry of Health and the Bureau of Tourism maintained that SARS epidemic in China was well under control and that the situation was safe for foreign tourists and visitors.[17] The cover-up of epidemic conditions led to complacency and risky behavior of the population and even medical staff, leading to the spread of SARS to Hong Kong, Southeast Asia, and North America.

3. Ineffectual coordination

Inter-agency and inter-regional coordination was ineffectual. Li Liming, the head of the national CDC, admitted at a domestic press conference that China's "medical departments and the mass media" lacked proper coordination.[18] The valuable lessons for containing SARS which Guangdong had learned through the deaths of medical staff were not passed on soon enough to other provinces badly hit by SARS, including Beijing.

The SARS management leadership small group headed by the Ministry of Health apparently lacked the authority to coordinate anti-epidemic work among ministries and bureaus which were at the same or higher bureaucratic rank. Moreover, in Beijing the military apparently did not report the full SARS statistics soon enough to the Ministry of Health.[19] Moreover, government and the health authority in mainland China and Guangdong failed to share the up-to-date and reliable SARS conditions and effective treatment with their counterparts in Hong Kong. This caused the rapid and costly spread of SARS in Hong Kong.[20]

4. Gaps in the law

The Law of Prevention and Treatment of Epidemics contained gaps. It only mandated the immediate upward reports of four epidemics, namely, plague, cholera, AIDS, and anthrax by medical units and personnel or by CDCs to the local and national health departments (Article 21). SARS, a new but highly contagious and deadly epidemic, was not listed. At the press conference on the SARS epidemic in Guangdong on February 11, 2003, the Director of the GDBH explained that the bureau did not to publicize the SARS epidemic quickly despite that Heyuan reported to it on the disease on January 2, because the Epidemic Law of 1989 did not stipulated it as a disease that should be publicized immediately.[21]

Epidemic management after April 2003

In response to the ineffectual SARS management in the early 2003, the Chinese government moved to overhaul epidemic management. The existing laws and regulations have been mended and new regulations have been introduced in order to fix the loopholes in these documents. In particular, two developments seem of particular importance. First, on August 28, 2004 the Epidemic Law of 1989 was amended. Second, on February 26, 2006 the Preparatory National Plan for Emergency Responses during Sudden Public Health Events (the Epidemic Crisis Plan for short) was announced by the State Council. The latter signified that epidemic management institutions at the national level were being formally established. In the following paragraphs, the epidemic management institutions will be outlined. This section analyses the main aspects of the nationwide epidemic management institutions after April 2003. While the plan and the Epidemic Law supply the most comprehensive rules regarding epidemic management, actual practice will also be examined.

Laws, regulations, and policy documents

The main policy and legal documents governing China's epidemic management include the following: the Epidemic Crisis Plan, the Epidemic Law amended in August 2004; the Law on Health and Quarantine within the National Boundary promulgated in May 1987; Regulations on Emergency Responses During Sudden Public Health Events (the Health Crisis Regulations for short) promulgated in May 2003; and Regulations on Health and Quarantine Related to Domestic Transport promulgated in March 1999. The Epidemic Crisis Plan classifies major categories of epidemic crises, outlines the national and provincial epidemic management agencies, defines the management responsibilities, and stipulates norms for epidemic reporting and management.[22]

Classification of health crises and epidemics

Four categories of health crises are defined in the epidemic crisis plan: (I) grave (*tebie zhongda*); (II) major (*zhongda*); (III) big (*jiaoda*); and (IV) ordinary (*yiban*). Class I is clearly defined in the plan to include the spread of plague or anthrax in and beyond a large or medium-sized city, or in more than two provinces; spread of SARS or human-infected bird flu; spread of any unexplained epidemic in more than two provinces; surface and spread of a new epidemic or an once-distinct epidemic; loss of samples of highly contagious pathogen or viruses; imports of highly contagious epidemics from countries or areas in communications with China; and any other grave epidemic identified by the Ministry of Health. In the amended Epidemic Law, Class A epidemics still include only plague and cholera, while SARS, humanly contagious bird flu, and snail fever (schistosomiasis) were newly added to the list of Class B. Now the grave epidemic crises are defined more broadly than they were in the 1989 Epidemic Law, allowing the national government to play a more active role in management of severe, unknown, or new epidemics.

Epidemic management agencies and funding

According to the Epidemic Crisis Plan, the administration remains the key governmental branch directly in charge of coordinating epidemic management, and the health department the key agency in charge of daily epidemic management. At the national level, in cases of need the Ministry of Health can propose to set up Headquarters for Emergency Responses During Sudden National Public Health Events (*quanguo tufa gonggong weisheng shijian yingji zhihuibu*). According to the Health Crisis Regulations, the Headquarters consist of representatives of relevant ministries, bureaus, and offices under the State Council as well as departments of the military, and the Premier would automatically become the General Commander. According to the Epidemic Crisis Plan, the people's government (the administrative branch) of provinces, cities (prefectures), and counties, as well as the military and the armed police should set up their health crisis command centers and expert committees equivalent to the national ones (see below for their descriptions). According to the Health Crisis Regulations, in cases of need the local governments can form headquarters for health crises, which is led by the head of the government (i.e., governor, mayor, or magistrate).

The Office for Health Emergency Responses (*weisheng yingji bangongshi*), which is also the Command Center for Emergency Responses During Sudden Public Health Events (*tufa gonggong weisheng shijian yingji zhihui zhongxin*). This office was installed at the Ministry of Health as a bureau-level agency back in April 2004, two years before the national epidemic management institutions were established. It is in charge of managing daily health crises at the national level. The office is assisted by the Expert Consultative Committee (*zhuangjia zixun weiyuanhui*), as well as medical agencies, CDCs, health supervisory agencies, and customs health inspection and quarantine agencies.

Like the epidemic management institutions prior to April 2003, the health department is responsible for organizing medical agencies, CDCs, and health supervisory agencies to investigate and management heath crises, for organizing the expert consultative committee, for declaring the class of the health crisis, and for inspecting the supervising the health crisis management. It is required to build a medical team to respond during epidemics. The health department is also in charge of enforcing health laws and regulations.

The medical units are responsible for receiving, transporting, and treating patients, managing and treating medical wastes and summarizing clinical analyses of epidemics. The CDC plays a key role in the prevention and control of epidemics, for testing and monitoring epidemics, and for direct responses to epidemics. The customs inspection and quarantine agency is responsible for epidemic management at the border checkpoints. The government has budgetary funds to subsidize the professional and technical agencies for managing health crises.

The epidemic management institutions at the national level are replicated at the local levels. For example, back in December 2004, the same year when the Office for Health Emergency Responses Office of the Ministry of Health was installed, the Office for Health Emergency Response at the GDBH (OHERGD) was in operation. In addition, the OHERGD was assisted by a large expert committee comprising 123 experts and by eleven medical expert groups from across the province.[23] Later, in May 2007 the Office for Emergency Responses at the Guangdong Provincial People's Government (the administration) (OERGD) was established. In December of that year, Guangdong promulgated its Preparatory Plan for Emergency Responses During Sudden Public Health Events (the Guangdong Epidemic Crisis Plan for short). Similar to the national plan it classified health crises into four levels, placed the main responsibility for epidemic management on the shoulders of the provincial administration, and suggested principles and measures for epidemic management that were clearly patterned after those in the national plan. The OERGD and the OHERGD have remained the key coordinating bodies for managing crises in general and health crises including epidemics in particular.[24]

It goes without saying that the Party Committee and the Party Secretary at each level remain the ultimate power center behind epidemic management. Therefore, they can also intervene by issuing policy guidelines or instructions. After April 2003, having learned a bitter lesson from SARS, the trend seems to be that they have paid closer and earlier attention to any emerging epidemic.

It is also worth noting that from April 2003 onward, great efforts have been made to enforce official responsibility on epidemic management. For example, when the nation or a locality is switched to a mode to prevent and contain a spreading epidemic, officials (especially administrative and health) will be held responsible for epidemic management in their jurisdiction. During the SARS campaign, from April to June 2003 thousands of officials lost their posts due to their ineffectual performance or negligent acts in containing SARS.

Prevention stage (planning for epidemic crises)

Ministries, Bureaus, and Offices under the State Council and local people's government at or above the county level are required by the Epidemic Crisis Plan to formulate epidemic crisis preparatory plans in light of their own conditions, the national plan, and the Health Crisis Regulations. The amended Epidemic Law requires the local governments to file their plan to the next higher level (Article 20). According to the Health Crisis Regulations (Article 11), the plan should include the following:

1 the membership and responsibilities of crisis management headquarters;
2 surveillance and early warning of crises;
3 the collection, analyses, reports, and announcement of crises;
4 agencies with expertise for crisis management and monitoring and their tasks;
5 classification of crises and plans for their management;
6 prevention and on-the-spot management of crises; emergency facilities, equipment, medicine, and instruments, as well as reservation and allocation of other materials and technologies;
7 building and training of the crisis management team.

As stated, in December, Guangdong promulgated the Guangdong Epidemic Crisis Plan, detailing agencies in charge, categories of epidemics, and basic measures for epidemic managements. The provincial authority formulated 110 public health crisis plans to cope with a great variety of health contingencies, including epidemics.[25]

Moreover, the local governments are required by the Health Crisis Regulations to make efforts to prevent epidemics, educate the public about epidemic management, set up their monitoring and early warning system, and ensure the supplies of equipment and materials for epidemics, and conduct regular drills of epidemic management. The cities with districts are required to establish a special epidemic hospital or designate a capable medical agency to take charge of treating epidemics. It is clear that in the wake of SARS the Chinese authority has paid much greater attention to the prevention of epidemics. As a result, in anticipation of any known epidemics more efforts have been made at this stage than before.

The Chinese national and local governments have stepped up investment in building the CDCs nationwide. By 2004, 6.8 billion yuan had been invested for that purpose. Out of this sum, 634 million yuan was arranged by the National Development and Reform Commission for the first-stage construction project of the National CDC. The construction was inaugurated in 2004, and aimed for 76,000 square meters of floor area.[26] Beefing up the epidemic diagnosis, detection, and monitoring capacity helps China to stem the rise and spread of epidemics.

From 2003 to 2006 the State Council pushed for the building of medical emergency centers nationwide. In centrally administered municipalities, provincial capitals, and prefecture-level cities an existing hospital was selected for upgrade and expansion and to be turned into a designated hospital for treating epidemic patients. In each county a county-level hospital was slated to build an epidemic department or ward. These hospitals nationwide will be the main medical network for directly handling epidemic patients in order to minimize infection. National and local governments each contributed to half of 11.4 billion yuan fund for helping central and western regions to build up the epidemic medical treatment system. Reportedly, by 2003, epidemic hospitals at the prefecture level were already in operation.[27]

In managing epidemics after April 2003 both the Party and administration have swung into action much earlier than they did for SARS. China's prevention of influenza H1N1 serves as a good example of the operation of the post-SARS overhauled epidemic management institutions. Influenza H1N1 originated in Mexico in early April 2009. By late April the epidemic had attracted international attention. On May 10, China noticed the first imported H1N1 case. The Chinese authority acted more decisively and much earlier at the prevention stage in order to stem the emergence and spread of this infectious epidemic. On April 28 and May 11, President Hu instructed the nation to guard against the flu. On April 30 the Politburo Standing Committee met to discuss the containment of the flu. The State Council meeting on April 30 announced the following measures to prevent H1N1 viruses: (a) cooperation would be initiated with Hong Kong, Macao, and the international community; (b) the Ministries of Health and Agriculture and General Administration of Quality Supervision would collaborate to prevent the epidemic; (c) strict health inspection was to be imposed at the border checkpoints, especially people from the affected regions, and the Law on Health and Quarantine within the National Boundary was to be activated; (d) travel health advisory notices would be issued; (e) cases and unexplained pneumonia and influenza should be monitored closely, cases should be promptly discovered and reported directly and early, and diagnoses, quarantine, and treatment of patients should be provided; (f) live pigs and pork products should be inspected closely; (g) the stock of anti-epidemic equipment, drug, and medical instruments should be built up; (h) the knowledge on prevention and control of the influenza should be popularized.[28]

In addition, the government has invested heavily in vaccination in order to stem the spread of the epidemic among the population. The government allocated over 1 billion yuan of special funds for research and development in vaccination for influenza A. In June 2009 the Ministry of Health received the H1N1 viruses from the WHO. By early September the vaccination for influenza A H1N1 was approved after clinical tests and started to be manufactured in China, the first nation in the world to do so. By January 2010, 59 million Chinese had received the vaccination, whereas the chance for cases of severe side-effects was merely 1.2 out of one million, comparable to the world level.[29]

Some of China's forceful measures in controlling influenza A were questioned by Mexico officials and the US press. The Mexican government accused China of discriminating against Mexican nationals without symptoms, including those who did not go to Mexico for months. For this reason it chartered a plane in May to fly some of its citizens back home. It was reported by *Washington Post* that international travelers with slight fever or in a flight where a passenger had a fever were quarantined and tested for influenza A.[30]

Monitoring of epidemics

The Health Crisis Regulations contain the following specific requirement for upward reporting of epidemics. Most of these stipulated timelines for epidemic reporting are much stringent than those in the Implementing Measures of the Epidemic Law. The Measures mandated reports in the cities of any outbreak of Class A epidemics, AIDS, and anthrax within six hours, Class B epidemics within 12 hours, and Class C epidemics within 24 hours (Article 35). It also required provincial health bureaus to report any outbreak of Class A epidemics or mass epidemics to the Ministry of Health within six hours.

1 (Article 19) Provinces are required to report within one hour to the Ministry of Health in the case of an outbreak or a likely outbreak of an epidemic, an unexplained disease affecting a group of people, and mass poisoning, as well as a loss of an epidemic virus sample or a poisonous agent.
2 (Article 20) Epidemic monitoring agencies, medical units, and other work units are required to report within two hours to the health department of the local county government any instance described in Article 19, and the recipient health department should report within two hours to the local government, the health department of the next higher level, and the Ministry of Health.
3 (Article 23) The health department of the province where the epidemic crisis occurs should report timely to its counterpart of the neighboring province. When necessary, the recipient health department should report timely to medical agencies in the province.
4 (Article 24) The government publicizes telephone numbers for individuals to report risks for public health, as well as negligence of governmental agencies or officials in responding to epidemics.
5 (Article 25) The Ministry of Health is responsible for publicizing information on health crises and can authorize provinces to publicize health crises in the areas under their jurisdiction. The information should be timely, correct, and balanced.

Furthermore, the amended Epidemic Law (Article 30) specifically requires the medical agencies in the military to report outbreak of epidemics to the Ministry of Health. This stipulation removes the loophole in the 1989 Epidemic Law.

Again, influenza H1N1 (influenza A) serves as a useful example. Not only greater efforts were made from the start to beef up the monitoring system, but also the disclosure and inter-regional sharing of epidemic information was more forthcoming. From mid April, the Ministry of Health had started to monitor the epidemic and activate the health crisis management institutions. The Ministry of Finance had also allocated 301 million yuan to support the monitor system of influenza A. The national budgetary support helped expand the number of laboratories in the system from 63 to 411 and that of designated hospitals from 197 to 556. The Ministry of Health also received budgetary allocation of 63 million yuan for detecting the viruses and providing diagnoses of the influenza. The Ministry of Agriculture also received 31 million yuan for monitoring the epidemic among the animal population and developing vaccination.[31]

In addition, epidemic disclosure had become much more efficient and transparent. On the morning of May 10, 2009, an influenza patient, a student who studied in the United States and visited his family in Sichuan, was found to be weakly positive in the two H1N1 virus tests. The medical experts in Sichuan met to discuss the case and agreed that he constituted a suspect influenza H1N1 patient. On the afternoon of the same day of this medical conclusion, the Health Bureau of Sichuan Province reported the case to the Ministry of Health and the case became the first case of influenza H1N1 in China. On the evening of the same day, the ministry announced the case to the nation.[32]

Chengdu Municipality immediately held a press conference at 3 a.m. on May 11 in order to head-off public panic, setting a rare precedent in press conferences in the early morning in China. Furthermore, inter-regional sharing of epidemic information was much improved. For example, the Bureau of Health in Beijing Municipality learned a great deal about the influenza through the nationwide epidemic report system which had information on China's first case of influenza A in Chengdu.[33]

Containment stage (crisis management measures)

The national government and the Ministry of Health are responsible for managing grave epidemics, and epidemics below that level fall on the shoulders of local governments. For a typical health crisis, the Health Crisis Plan (Article 4.2) clearly prescribes the following measures for local governments:

1. Organize relevant departments to join the crisis management.
2. Mobilize personnel, materials, transporting vehicles, and relevant facilities and equipment.
3. Define an area for disease control. When a Class A or Class B epidemic takes place, a local government at the county level or above can declare the epidemic area. Provinces can blockade areas infected by a Class A epidemic under their jurisdiction. However, decisions on blockading large or medium-sized cities, epidemic areas across provinces, major transport routes, or national borders have to be made by the State Council.
4. Adopt dramatic measures to contain epidemics, including restrictions on or closure of fairs, assemblies, cinema shows, factories, business, and schools; restrictions or a halt to massed gathering activities; temporary requisition of houses and transport vehicles; closure of contaminated drinking sources.

5 Manage the migrant population. Introduce epidemic prevention and control of migrants, and quarantine and treat patients.
6 Introduce health inspection at transport hubs and border checkpoints.
7 Promptly release correct and balanced information.
8 Organize lower-level governments to collect information and dissipate and quarantine relevant people.
9 Ensure supplies of commodities, suppress rumors, and crack down on price gauging and swindling.

Again, a case of influenza A can be informative. It was spreading around the world from April 2009 onward. The first imported case of the influenza into China was reported on May 10. Premier Wen held three executive meetings of the State Council to arrange for the prevention of the disease, on April 28, May 5, and May 11, respectively. Many of the aforementioned measures prescribed in the meeting on April 28 remained in place.[34] On April 29, 2009, the Ministry of Health issued a draft technical guide on controlling the swine flu. The next day it renamed the swine flu as influenza A H1N1 and put it under Class B epidemics. On May 12, 2009 it promulgated the medical treatment program for the flu.[35]

Upon discovering the first suspect case of influenza H1N1 on May 10, 2009, the Health Bureau of the Chengdu Municipality immediately switched the focus from prevention to containment of the epidemic. It announced four measures at a press conference on May 11: (1) stepping up cooperation with the customs inspection and quarantine agencies and preventing additional "imported" cases of the influenza; (2) monitoring potential subsequent cases from the known case and minimizing the spread of the epidemic; (3) notifying the designated hospitals and fever clinics in the province to promptly report and check the cases of unknown pneumonia and influenza; (4) publicizing the epidemic and needed measures.[36]

In the following months, Sichuan Province adopted a number of major measures to contain influenza A. They included vaccination of 1.29 million people, especially the those who were vulnerable or were likely to be exposed to infection in their posts, arranging different work shifts for employees of governmental agencies and public institutions to reduce congestion and infections, banning of mass assemblies at the schools in Chengdu, and daily reports of the epidemic conditions by the schools and universities during the weekdays to the Chengdu Municipal Government.[37] With these measures in place the majority of residents in Chengdu surveyed were confident that the government would be able to contain the epidemic.[38]

Animal-related epidemic management

The aim of this chapter is to examine human epidemics. Nevertheless, it is also worthwhile to briefly outline the animal epidemic management, as certain animal epidemics can be humanly contagious, such as influenza H1N1 and bird flu.

On February 27, 2006 the State Council promulgated the Preparatory Plan for Managing Major National Animal Epidemics (Animal Epidemic Crisis Plan for short). Like the epidemic crisis plan, the animal epidemic crisis plan classified four levels of animal epidemic crises. It also specified agencies and headquarters managing animal epidemics, as well as monitoring and early warning, and reporting of and principles for managing animal epidemics. In addition to the plan, a number of laws and regulations help define the animal epidemic management institutions – the Law on Prevention of Animal Disease; the Law on Health Inspection of Imported and Exported Animals; and the Regulations on Emergency Responses During Major Animal Epidemics.

The Ministry of Agriculture is the primary national agency in charge of major national animal epidemics. In cases of need during major epidemics, national headquarters can be formed, and the director of the General Office of the State Council and the Minister of Agriculture serve as their head and deputy head, respectively. Major provincial animal epidemics should be managed by headquarters headed by the governor, assisted by local veterinary administrative departments. Local veterinary administrative departments (usually the Bureau of Agriculture) are responsible for local animal epidemics. Animal epidemic prevention agencies and customs health inspection agencies are professional health agencies responsible for surveillance, detection, quarantine, and other medical procedures for containing animal epidemics.[39]

Conclusion

Since 1991 China's epidemic institutions have included a legal framework and have involved the administration, the health department, CDCs, hospitals, and clinics. The law and relevant regulations also categorized epidemics and stipulated prevention, monitoring and reporting, and containment procedures. However, prior to April 2003, China's epidemic management was undermined by several institutional weaknesses – Party and administrative leaders' preoccupation with economic growth and social stability, lack of a preparatory plan and a practical guide for epidemic management, weak early warning and detection, inadequate financial input in the medical and anti-epidemic agencies, weak inter-regional and inter-agency coordination, slow disclosure of epidemic information to the public, and loopholes in the Epidemic Law and regulations. As a result, when the first SARS cases caught the attention of the local health department in Guangdong in early 2003, both local and national health departments and leaders responded to the epidemic in a secretive, belated, and uncoordinated manner. As a result, SARS rampaged in Guangdong and northern China, China's economy was temporarily damaged, and its international image was tarnished.

In mid April 2003, the Hu-Wen leadership acted decisively to manage SARS. It has overhauled epidemic management institutions since then. Information about epidemic conditions became transparent, inter-regional and inter-agency coordination was much improved, official accountability for epidemic management was enforced, and regulations on epidemic management were quickly promulgated.

In the subsequent months, the Epidemic Law was amended and several noticeable loopholes were remedied; a national epidemic management plan was announced, and local epidemic plan was mandated. As a result, epidemic management institutions at the national level have been established. A number of components of epidemic management are similar to those prior to April 2003. They included the agencies involved, their respective roles, and most of the measures prescribed in response to the epidemics stipulated in the previous anti-epidemic institutions or documents. Nevertheless, many significant improvements have been made. The Party and the administration have viewed proactive measures as beneficial to instead of detrimental to effective epidemic management and economic growth. Timely and transparent information on epidemics, especially during the monitoring stage, are instituted and the stipulation is largely followed. Institutional preparedness and coordination, as well as financial inputs, have been rightly recognized.

Influenza H1N1 (also known as swine flu) provided a glimpse of the operation of the newly established national epidemic management institutions. Overall, China reacted swiftly and decisively to the epidemic and has been able to minimize the health damage of the virus. Some of its heavy-handed measures, however, were criticized by countries like Mexico or the United States for excessively restricting patients' or suspect patients' rights. These restrictions on people's rights deem questionable in countries (such as the United States and United Kingdom) where personal freedom is extolled. But in China (and probably East Asian countries) with a collective culture and which has been long exposed to the devastation of pandemics, the public apparently cherishes a strong state's swift response to epidemics and quick success of controlling epidemic rampage. The Chinese public seemed to believe that these measures were justified given China's past painful experience with epidemics and its densely populated coastal and central regions. In contrast, the lax measures practiced in North America (and even Europe) seem to lead to a faster spread of the epidemic and a higher toll.

As a result of China's aggressive anti-epidemic efforts, the casualties of influenza A in China have remained relatively low. By January 10, 2010, there were 123,196 confirmed H1N1 cases in China, 113,994 of which were successfully cured.[40] The ratio of cases to the population was about one out of 10,000 and the death to population was roughly one out of 1.8 million. In contrast, according to the estimates by the CDC in the United States, influenza A was estimated to have infected about 60 million people in the United States, about 15 percent of the population and over 480 times higher than that of China.[41] Not surprisingly, it was reported that as high as 85 percent of the Chinese surveyed were satisfied with the government's efforts to control influenza A. Management of influenza A thus became the government's medical efforts with the highest public approval in 2009. The WHO also praised China's efforts to contain H1N1 viruses as highly effective, a marked departure from the outright criticism which this body leveled at China in early 2003 for its mismanagement of SARS.[42]

China's swift and aggressive response to the pandemic of influenza A suggest that once China activates its epidemic management institutions at national and local levels, it can effectively contain viruses and prevent the epidemic from claiming many lives, even though its forceful anti-epidemic efforts entail considerable costs for the freedom of suspect patients and people in possible contact of them.

One critical reason for the quick improvement in China is that once the Chinese state recognizes the importance of effective epidemic management, it can overhaul the flawed institutions. It is also capable of mobilizing its bureaucracy, medical units, and even the population to fight epidemics.

In some ways it is true that the non-democratic nature of the regime may undermine epidemic management, especially when the leaders do not recognize its importance and are preoccupied with other tasks and when epidemic information is held secret. However, once the Chinese authority values the usefulness of swift epidemic management its leaders can act promptly to address the outstanding flaws in the epidemic management institutions, and can discipline the bureaucracy and build the medical system. As a result, they can be transformed into effective fighters against epidemics.

Notes

1 For literature on SARS in China, refer to John Wong and Zheng Yongnian (eds), *SARS Epidemic: Challenges to China's Crisis Management* (Singapore: World Scientific, 2004); Arthur Kleinman and James Watson (eds), *SARS in China: Prelude to Pandemic?* (Stanford, CA: Stanford University, 2006); Tommy Koh, Aileen Plant, and Eng Hin Lee (eds), *The New Global Threat: Severe Acute Respiratory Syndrome and Its Impact* (Singapore: World Scientific, 2003). For a critical analysis of SARS management in Guangdong, refer to Lai Hongyi, "Local Management of SARS in China," in Wong and Zheng (eds), *SARS Epidemic*, pp. 77–98.
2 Zou Keyuan, "SARS and the Rule of Law in China," in Wong and Zheng (eds), *SARS Epidemic*, pp. 101–102.
3 For a text of the Epidemic Law of 1989, refer to Xue Lan, Zhang Qiang, and Zhong Kaibin, *Weiji guanli* [*Crisis Management in China*] (Beijing: Qinghua daxue chubanshe, 2003), pp. 348–354.
4 Wang Ke'an, "Zhongguo mianyi shiye de chengjiu yu zhanwang [Achievements and Prospects of the Immunisation Project in China]," *Zhongguo liuxingbingxue zazhi* [*Chinese Journal of Epidemiology*], Vol. 20, No. 6 (1999), posted at http://www.yufangz.com/xgfg/ylun/200605/214.htm (accessed on July 24, 2010).
5 "Aiguo weisheng yundong [Patriotic Sanitary Campaign]," posted at http://baike.baidu.com/view/192267.htm (accessed on July 24, 2010).
6 Zou Keyuan, "SARS and the Rule of Law in China," pp. 101–102.
7 Xu Nantie, *Fei dian de dianxing baogao* [*A Typical Report on "Atypical Pneumonia"*] (Nanhai: Guangdong renmin chubanshe, 2003), pp. 4–30.
8 John Pomfret, "China's Slow Reaction to Fast-Moving Illness," *Washington Post*, April 3, 2003, Page A18.
9 Xu Nantie, *Fei dian de dianxing baogao*, pp. 49–50.
10 Ibid., pp. 54–62.
11 Ibid., pp. 71–74.
12 Ibid., p. 78.
13 Ibid., pp. 87–90; Yang Chunnan, Xu Qingyang, and Xiao Wenfeng, "Zujizhan: linwei buluan, chenzhe yingzhan [Blocking Action: In the Hours of Danger, Avoid

Confusion and Meet the Attack Calmly]," *Xinhua*, April 22, 2003, posted at http://news.xinhuanet.com/focus (accessed on April 22, 2003). See Caijing zaizhi bianjibu [Editorial Department of Finance and Economics Magazine], *SARS diaocha tekan* [*Special Edition of SARS Investigation*] (Beijing: Zhongguo shehui kexue chubanshe, 2003), pp. 156–159.

14 John Pomfret, "China's Slow Reaction to Fast-Moving Illness: Fearing Loss of Control, Beijing Stonewalled," *Washington Post,* April 3, 2003. For a detailed analysis of Guangdong's management of SARS, refer to Tony Saich, "Is SARS China's Chernobyl or Much Ado about Nothing," in Kleinman and Watson (eds), *SARS in China*, pp. 74-78; Joan Kaufman, "SARS and China's Health-Care Response: Better to Be Both Red and Expert," in Kleinman and Watson (eds), *SARS in China*, pp. 55–56.

15 For literature on SARS in China, refer to Note 1.

16 For the negative effect of this political concern on detection and treatment of the SARS in the national CDD, see Cao Cong, "'Chinese Scientists Were Defeated by SARS'", in Wong and Zheng (eds), *SARS Epidemic*, pp. 157–180.

17 This point was made in Tony Saich, "Is SARS China's Chernobyl or Much Ado about Nothing," p. 73.

18 Ibid., p. 80.

19 According to Saich, Premier Wen, on his visit to the national CDC, criticized the military for not reporting its SARS cases to the Ministry of Health (ibid., p. 81).

20 For an analysis of fragmented bureaucracy during the initial SARS management, see Lai Hongyi, "Local Management of SARS in China," pp. 87–90.

21 Xu, *Fei dianxing de dianxing baogao*, pp. 78–81.

22 These documents are collected in Zhongguo fazhi chubanshe (ed.), *Yingji guanli gongzuo shouce* [*Handbook for Emergency Management*] (Beijing: Zhongguo fazhi chubanshe, 2003), pp. 11–73.

23 "Guangdongsheng weishengting yingjiban jianbao [Briefings of the Office of Emergency Response of the Bureau of Health of Guangdong Province]," No. 1 and No. 4 (2004), posted at http://www.gdwst.gov.cn/a/wsjb (accessed on July 25, 2010).

24 "Guangdongsheng tufa gonggong weisheng shijian yingji yu'an [Preparatory Plan of Guangdong Province for Emergency Responses During Sudden Public Health Events]," announced by the Office for Emergency Responses at the People's Government of Guangdong Province, December 2, 2007, posted at http://www.gdemo.gov.cn/yasz/yjya/zxya/ggwslya/200712/t20071202_36148.htm (accessed on July 25, 2010).

25 "Guangdongsheng weishengting yingjiban jianbao."

26 "Jianquan tufa gonggong weisheng shijian yingji jizhi gongzuo de baogao [Report on the Work of Consolidating the Mechanism to Cope with Public Health Incidents]," *Zhongguo rendaxinwen* [*National People's Congress of People's Republic of China Newspaper*], April 15, 2004, posted at http://health.sohu.com/2004/04/15/22/article219842254.shtml (accessed on July 25, 2010).

27 Ibid.

28 "Wen Jiabao zhuchi Guowuyuan huiyi, bushu zhuliugan fangkong gongzuo [Wen Jiabao Chaired a State Council Meeting and Arranged Prevention and Control of Swine Flu]," *Xinhua*, April 28, 2009, posted at http://news.qq.com/a/20090428/001205.htm (accessed on April 21, 2010).

29 "Zhongguo chao bacheng gongzong manyi zhengfu fangkong jialiu ju cuo [Over Eighty Percent of the Public in China Satisfied with Governmental Measures for Preventing and Controlling Influenza A]," *Guangming Daily*, January 15, 2010, posted at http://www.39world.com/healthnews/0123T94O2010849479.html (accessed on April 21, 2010).

30 Ariana Eunjung Cha, "Caught in China's Aggressive Swine Flu Net: Quarantine Measures Keep Cases Down but Virtually Imprison Healthy Travelers," *Washington Post,* May 29, 2009.

31 "Duikang jialiu: woguo zhangfu caiqu de cuoshi [Countering Influenza A: Measures Taken by the Government in Our Country]," posted at http://13058757.blog.hexun.com/40077804_d.html on November 10, 2009 (accessed on April 21, 2010).

32 "Woguo neidi faxian shouli jiaxing H1N1 liugan yisi bingli [The First Suspect Case of Influenza H1N1 Is Found in Mainland of Our Country]," posted at http://bbs.xxrb.com.cn/thread-99665-1-1.html (accessed on July 29, 2010).

33 "Duikang jialiu: woguo zhangfu caiqu de cuoshi."

34 "Wen Jiabao zhuchi Guowuyuan huiyi, bushu zhuliugan fangkong gongzuo."

35 "Duikang jialiu."

36 "Fasheng shuruxing bingli, wosheng you 'fang' zhuan 'kong' [As an Imported Case of Disease Occurred, Our Province Has Switched from 'Prevention' to 'Control']," announced at Chengdu Municipal People's Government Press Conference, May 12, 2009, posted at http://www.chengdu.gov.cn/newsrelease/list.jsp?id=256092 (accessed on July 29, 2010).

37 "Fang jialiu kuoshan, jiguan shiye danwei cuoshi shangban [In Order to Prevent the Spread of Influenza A Agencies and Institutions Introduce Different Shifts]," *Huaxi dushi bao* [*West China City News*], September 12, 2009, posted at http://www.wccdaily.com.cn/epaper/hxdsb/html/2009-09/12/content_92530.htm (accessed on July 29, 2010); "Shengweishengting zuori zhaokai fabuhui, 129 wan ren jiezhong jialiu yimiao [Provincial Bureau of Health Convened Press Conference Yesterday and 1290 Thousands of People Received Influenza A Vaccination]," December 30, 2009, posted at http://cd.goodcar.cn/09/1230/000074339.shtml (accessed on July 29, 2010); "Chengdu fangkong jialiu, shixing 'ri baogao zhidu' [In order to Prevent and Control Influenza A, Chengdu Introduces "Daily Report" Institutions]," September 8, 2009, posted at http://www.sina.com.cn (accessed on July 29, 2010).

38 Han Li, "Shiming "Fangkong jialiu" xinxing zhishu gaoda 94.8% [Confidence Index of City Residents in 'Preventing and Controlling Influenza A' Reached High 94.8%]," *Chengdu shangbao* [*Chengdu Business Times*], November 19, 2009, posted at http://www.chengdu.gov.cn/glamor_chengdu/detail.jsp?id=291789 (accessed on July 29, 2010).

39 See "Guojia tufa zhongda dongwu yiqing yingji yu'an [The Preparatory Plan for Managing Major National Animal Epidemics]," in Zhongguo fazhi chubanshe (ed.) *Yingji guanli gongzuo shouce*, pp. 169–184.

40 "Zhongguo chao bacheng gongzong manyi zhengfu fangkong jialiu ju cuo."

41 Data for influenza A in the United States by March 13, 2010 came from the following source: Centers for Disease Control and Prevention (of the United States), "CDC Estimates of 2009 H1N1 Influenza Cases, Hospitalizations and Deaths in the United States, April 2009 – March 13, 2010," a document of Centers for Disease Control and Prevention, April 19, 2010, posted at http://www.cdc.gov/H1N1flu/pdf/H1N1_Estimates_Apr19.pdf (accessed on April 24, 2010). Earlier data come from the following source: "ECDC Daily Update: Pandemic (H1N1) 2009, 18 January 2010," a document of the European Centre for Disease Prevention and Control, January 18, 2010, posted at http://ecdc.europa.eu/en/healthtopics/Documents/100118_Influenza_AH1N1_Situation_Report_0900hrs.pdf (accessed on April 23, 2010).

42 "Zhongguo chao bacheng gongzong manyi zhengfu fangkong jialiu ju cuo."

6 China's management of environmental crises

Risks, recreancy, and response

Richard P. Suttmeier

Introduction – a country at risk

As the crisis events of the last several years remind us, China is a nation of risks. Serious floods, deadly landslides, chemical spills, industrial accidents, drought, and catastrophic earthquakes all point to China's vulnerability to natural and human-created hazards. In an average year, thousands of people die as result of natural hazards and millions more are otherwise affected by injury and the loss of homes and property.[1] Environmental pollution and industrial accidents kill and maim additional thousands. The risks are diverse, are widely distributed around the country, occur with relatively high frequency, and typically involve high losses.[2] Ongoing urbanization, increasing inter-regional interdependencies, and the deployment of more demanding and powerful technologies increase the likelihood that the magnitude of catastrophic incidents will increase. The management of crises resulting from these risks poses serious governance challenges for the Party-state. It must both devise effective emergency responses to the multiple crises which a society of risks engenders while also meeting the increasingly pressing need for risk-reduction measures, a challenge which ultimately involves a reconfiguration of incentives and patterns of responsibility in both state and society.[3]

Many of today's risks are certainly not new to China. Chinese history is punctuated by major earthquakes, for instance, with four in particular taking more than 200,000 lives each.[4] Major floods, typhoons, famines, droughts, and epidemics have been more frequent occurrences. The traditional state was conscious of these risks and devised crisis management policies and institutions to respond. Given China's size and complexity, and given the ubiquity of its bureaucratic system, it is not surprising that central tenets of this crisis management tradition involved establishing a hierarchy of incidents in terms of their severity, with efforts to parcel out responsibilities for them between central and local authorities, depending upon severity. At the same time, China's traditional approach to crisis management also recognized the limits of the state and the importance of mobilizing local elites outside of the state for both disaster prevention and crisis response.[5] While remarkable in a variety of ways, the traditional crisis management regime was nevertheless subject to failure, typically by falling victim periodically to debilitating recreancy[6] and a breakdown of a capacity for collective action. As in other countries, the failure of crisis management capabilities reflected negatively on political legitimacy and, in extreme cases, led to the loss of the "mandate of heaven."

The Communist regime understood well the relationships between effective crisis management and political legitimacy when it took power in 1949, and strove to both protect Chinese citizens from risks and to devise response mechanisms when disaster struck. Unfortunately, the developmental trajectory of Maoist China included innumerable misguided policies and practices which in many ways put the country at greater risk, especially when considering the increased possibilities for damage resulting from industrialization, population growth, and the deployment of more powerful technologies.[7] At the same time, its approach to the development of crisis response mechanisms resisted modernization, relying largely on ex post-facto mass mobilizations and relatively weak crisis management planning.[8]

The China of the late Mao era, thus, had become one which had entered a "risk transition," where new modern risks resulting from industrialization, population growth, and urbanization existed side-by-side with traditional risks, and where the crisis management mechanisms had yet to be adapted to the new risk realities.[9] The initiation of the programs of reform and openness in 1978 accelerated the movement into the risk transition phase as China's risk profile changed in a variety of ways. Rapid economic growth has been accompanied by dreadful environmental degradation and severe ecological stresses which frequently produce crisis incidents, and the introduction of many new and powerful technologies, somewhat Janus-faced, both exacerbate environmental and safety problems but also offer new opportunities for more effective risk management.

China's increasingly complex and deep engagement with the international community has also exposed it to new norms of risk management and international best practices for responding to crisis incidents. At the same time, the years of marketizing reform has also had the effect of removing many of the "guarantees" that constituted the socialist safety net Chinese citizens had taken for granted. Thus, Chinese citizens have experienced a psychological shift characterized by heightened awareness of risk – economic, safety, environmental – in society. The old "total insurance" assumptions of socialist society have given way to new uncertainties. In spite of these developments, China had, until recently, been slow to develop new institutions for managing risks and responding to crisis situations. Thus, Aaron Wildavsky's observation that "richer is safer" has yet to apply unambiguously to China.[10]

Risk, emergency, and crisis

In attempting to understand China's current approach to crisis management in the areas of natural disasters, environmental pollution, and industrial accidents, it is helpful to reflect on the evolution of thinking in the international community on these issues. As China has struggled to develop an improved crisis management system, it has consciously sought to study, and where appropriate, emulate international experience. This has occurred through its interactions with international organizations with active crisis management programs, especially United Nations agencies,[11] through regional and bilateral programs of cooperation, and increasingly through the development of a professional crisis management research tradition in China which has studied crisis management approaches in other countries.

However, there are inherent limitations to the relevance of international experience. Crisis incidents inevitably involve many distinctive local conditions, and local responses to crisis incidents shape the character and effectiveness of emergency response. More importantly, the development of a crisis management system reflects not only the influences of international best practices, but also the path dependency of national and local institutions. Thus, it is useful to see China's approach to crisis management as involving a tension between international norms and the values and traditions of its extant institutions. In this sense, China's approach to crisis management parallels its experience with reform and development over the past several decades more generally. Institutional models from the international community have been studied and selectively introduced. Advanced technologies have been imported and deployed. But, through it all, certain core values and practices of political institutions have persisted, and/or resisted reform. As with China's broader development experience, its evolving approach to crisis management raises the question of whether China's relative ability to accommodate both global influences, and local and national institutional interests, is best understood either as the creation of a new twenty-first-century "Chinese model" of governance, or as the source of potential societal instability.

The concepts of *risk* and *risk management* have come to figure prominently in international thinking about crisis management. As catastrophic incidents have increased around the world, the sense of being "at risk" has become more widespread, and with it, increased concern for the costs of disasters and ways of reducing the costs of catastrophic incidents. Growing professional attention to the management of crises has been accompanied by interest in securing the best returns from increased investments in crisis management. Wise investment decisions are those which normally take account of some sort of risk–benefit analysis, where risk is understood as the probability of an event times the magnitude of its impact.

Understanding risk in these terms points to a fundamental conundrum in building crisis management systems. Crisis incidents are usually understood in terms of low-probability but high-impact events, thus complicating the investment intervention decision. If an event is of low probability, investment in prevention or mitigation might be considered a poor use of scarce resources, especially in developing countries. On the other hand, the cost of crisis incidents can be staggering, with the result that interventions taken to lower those costs often can be more than justified. The questions then become, when are such investment interventions justified, and how are institutional arrangements put together to assess risk and make investment decisions?

Much recent research suggests that investments in "pre-event" activities warrant considerably more attention; crisis management thinking should therefore move beyond the immediate "emergency response" challenges to a more comprehensive risk management perspective. Crisis managers are therefore enjoined to think not only in terms of response mechanisms for immediate post-event actions, but also in terms of pre-event investments in prevention and mitigation activities, in the preparations for response, in preparation for immediate post-event recovery challenges, and in planning for the longer-term recovery.[12] Again, since it requires action by a variety of organizations across a variety of organizational jurisdictions, one of the more challenging tasks for crisis managers is to devise institutional arrangements that will permit the effective adoption of this more comprehensive risk management approach. As we will see, an understanding of the importance of this comprehensive approach is taking root in China, but progress towards its implementation has been quite mixed.

The overlapping of the concepts of "risk management," "emergency management," and "crisis management" is evident when a crisis is understood in terms of three main components – *threat*, *urgency*, and *uncertainty*. Crises are understood in the first instance when values, lives, and property are threatened, or are at risk. Of course, many threats (e.g., climate change) seem to be longer-term and do not elicit crisis responses. Crises, like emergencies, therefore typically involve the compression of time, or urgency, where responses from decision-makers must be made quickly. In many situations where lives and property are at risk, and where decisions must be made quickly, preplanned routines, characteristic of sound emergency management practices, can be invoked with some confidence that the threat is understood and that the consequences of response are reasonably well known. Crises, and the other hand, can be distinguished from emergency incidents by the levels of uncertainty surrounding the causes of the threat, the ways in which the incident unfolds or develops, and the likely consequences of intervention.[13]

In their recent work on crisis management, Arnold M. Howitt and Herman B. Leonard draw even finer distinctions between emergencies and crisis, arguing that the novelty associated with a crisis – resulting from the scale of an incident (e.g., a major earthquake, typhoon, or hurricane) and/or the strange, unfamiliar, or unexpected pattern of its development (e.g., a nuclear accident, a deep-sea oil-well blowout) – often renders established emergency procedures irrelevant or counterproductive. "Routine emergencies" (referred to as "Mode R" incidents) are therefore distinguishable from "crises" ("Mode C") by the high degree of uncertainty found in the latter. Mode C incidents thus put a premium on "sense-making" in the face of the unfamiliar, and on creative, adaptive decision-making that is not presumed in the scripted routines typically found in good emergency management procedures. Effective crisis management organizations must therefore be prepared to operate in both Mode R and Mode C, by no means an easy challenge given the different organizational requirements of the two modes.[14]

As we shall see, Chinese experience builds on emergency management traditions from the past and on recent disaster experience combined with lessons learned from the international community. Since a number of recent incidents have exhibited crisis characteristics, the challenge of developing crisis management capabilities has become pressing. Let us review several of these before examining recent developments of China's crisis management regime in greater detail.

A mini-Bhopal?

There is no shortage of environmental insults and industrial accidents which have put, and continue to put, Chinese citizens at risk. One particularly striking case, considering the ghoulish consequences, nicely illustrates many of the problems with China's crisis management system. At 9:55 p.m. on the night of December 23, 2003 in Kaixian County, Chongqing municipality, a natural-gas well experienced a serious blowout resulting in the release of a large amount of hydrogen sulfide. The release affected 28 villages in four towns, killing more than 200 people, either in their sleep, or as they attempted to escape the gas. Some 4,000 others were hospitalized.

The investigation that followed the incident found that the Eastern Sichuan Drilling Company, a subsidiary of China National Petroleum Corporation was negligent. The negligence of the company, which supposedly was using safe and reliable advanced technology, involved the removal of a critical blowout prevention valve, a failure to promptly ignite the gas (which would have resulted in the production of relatively harmless by-products) until 18 hours after the event, and a failure of the workers to promptly notify the local government, the obvious first responder. But, media accounts of the incident also illustrated that much was wrong with China's approach to crisis management more broadly in the face of potentially catastrophic incidents.

As a state-owned company, East Sichuan Drilling had a bureaucratic status that required that its regulatory oversight be provided by the municipal, rather than by the county government. As a result, the standard operating procedure dictated that the municipal government should be notified first in the event of an incident. As a result, it was only one-and-a-half hours after the initial release that the county government was notified of the accident by the municipal government, thus wasting critical time during which – in principle – lives could have been saved. Whether the county government could have done more, though, is also an open question, since it did not have the equipment to safely enter the affected area to rescue residents, nor did most residents have telephones which could have been used to alert them in a timely fashion. Thus, faulty decision-making mechanisms combined with weak emergency preparedness at the local level helped turn a controllable accident into a deadly crisis.

Further evidence of a flawed crisis management system are the indications of official recreancy, especially with reference to government–industry relations.

The government failed to require what should have been a mandatory environmental and safety assessment of the site, thus saving the company the cost of the evaluation (estimated to be in the range of $US 200,000–300,000) and also failed to enforce zoning regulations which prescribed that there be no residences within a one-kilometer-diameter zone around the well.[15]

The Kaixian case illustrates very nicely how crisis management measures had fallen well behind the changing risk profile of a China experiencing rapid economic development, new technologies, and population growth. Not only was the emergency response deeply flawed, but more importantly, the pre-event risk reduction and risk mitigation opportunities were tragically overlooked.

Intensifying crises, learning, and reform

Most observers of China's approach to crisis management today call attention to the impacts which the 2002–03 outbreak of Severe Acute Respiratory Syndrome (SARS) had on Chinese thinking. The SARS experience, involving a new and unconventional risk, revealed many problems with China's approach to managing crisis situations. These included problems of interagency coordination, coordination between local (especially provincial) and central governments, civil–military relations, international liaison, expert scientific advice, and public relations/information management. But, in addition, as a new and unfamiliar phenomenon, Chinese crisis managers failed in the "sense-making" task of understanding and characterizing the risk, thus contributing to the severity of its consequences.[16]

The serious problems of crisis management revealed in the SARS episode prompted the Chinese government to rethink its crisis management regime and initiate a number of important institutional changes. The most notable of these was the establishment in December, 2005, of a "one office, four committees" crisis management system. It includes a new National Emergency Management Office in the State Council (*guowuyuan yingji guanli bangongshi*), which is intended to achieve national coordination among the many agencies involved in crisis management, and see to the development of up-to-date contingency plans at both the national and local levels, and four more specialized interagency committees – the National Committees for Disaster Reduction, Work Safety, Food Safety, and the Communist Party's National Committee for Integrated Crisis Management (having responsibility, respectively for natural disasters, industrial accidents, public health and food and drug safety emergencies, and public security incidents).

The committees are led by a vice-premier with staff support coming from four specialized State-Council-level entities: the National Disaster Reduction Center of the Ministry of Civil Affairs (MOCA); the Supervision Center for Work Safety of the State Administration of Work Safety; the Emergency Management Office and Center for Disease Control and Prevention of the Ministry of Public Health; and the Committee for Comprehensive Management of Public Security of the Ministry of Public Security.[17] Specialized units of these sorts have been established at local government levels as well, with efforts made to achieve "unified leadership combining vertical and horizontal agencies."[18] In addition, at the national level, there are also special State Council "leading groups" dealing with emergency situations, such as the State Flood Control and Drought Relief Headquarters, and the State Earthquake Relief Headquarters.

The new emergency response system also calls for the development of contingency plans for responding to crisis incidents in which an incident is ranked and color-coded by the degree of severity. Four levels are specified, with lead responsibility being assigned to different levels of government depending on the nature and seriousness of the emergency (location and spatial dimensions, numbers killed, etc.). Level I incidents elicit the highest response, with a vice-premier taking the lead, Level II responses are led by a minister, etc. The hierarchical system of crisis incident classification is also used at local levels, where, in the event of a major Level I emergency, for instance, a provincial governor would be designated as commander.

Much is made in China about the importance of local government roles in crisis management, with the local government being expected to take the lead whenever possible.[19] Clearly, local emergency response mechanisms are quite important, since they are likely to result in the deployment of first responders. It is also clear that accurate information from the site of the incident is critical for activating the system both at national and local levels. This in turn calls attention to the importance of situational awareness on the part of local officials and a willingness to provide prompt and candid reports in the event of an emergency. But, as seen in the Kaixian case, and in countless others, including the Songhua River case, discussed below, these conditions have not always been met. As a result, controllable emergencies have escalated into major crises.

As part of the new crisis management system, the State Council also formulated a national Master Plan (*guojia tufa gonggong shijian zongti yingji yu'an*) which reflects the institutional changes noted above. Subordinate to the Master Plan are 25 incident specific "thematic" plans (*zhuanxiang yingji yu'an*) relating to floods, earthquakes, forest fires, industrial accidents, aviation accidents, ocean salvage and rescue, urban subway accidents, blackouts, nuclear accidents, environmental emergencies, telecommunication emergencies, public health crises, emergency medical assistance, animal disease outbreaks, food safety incidents, financial emergencies, and emergency incidents abroad, as well as 80 central departmental plans (*bumen yingji yu'an*) and local emergency plans, the most important of which are the provincial emergency plans (*shengji zongti yingji yu'an*).[20] In addition, in November, 2007, the National People's Congress enacted a new Emergency Response Law.

The November, 2005 chemical spill on the Songhua River further illustrated why the development of a modern and effective crisis management system was needed. The spill was caused by a powerful explosion which killed five people and wounded 70 at the Jilin Chemical Industrial Co., a subsidiary of PetroChina. The explosion then led to the release of 100 tons of benzene into the Songhua

River. Releases of information about the spill were delayed and coordination among relevant agencies at both national and local levels left much to be desired.[21] Indeed, it has been argued that the failure of the emergency management system again helped turn an emergency into a crisis. The explosion at the Jilin chemical plant which led to the release of the benzene occurred on November 13. But, officials in downstream Harbin only learned of the release on November 18, and they, in turn, did not turn off the city water supply until November 22.[22] The State Environmental Protection Administration (SEPA) (now, the Ministry of Environmental Protection, or MEP) was also very slow to react, and its poor performance led to the resignation of SEPA Minister Xie Zhenhua, and an invitation to the UN Environment Program to review the incident.[23]

The Songhua incident further revealed a number of familiar problems with the crisis management system, including problems of low emergency awareness, misaligned incentives for preventing crises, inadequate emergency preparedness, inadequate financing of crisis management activities, and problems of coordination, monitoring, and reporting. Coming as it did at roughly the same time that innovations in the emergency management regime were being introduced, it accelerated the planning process for environmental emergencies, as indicated by the promulgation of the "National Plan for Environmental Emergency Response" by the State Council in January, 2006, and the subsequent establishment of an Environmental Emergency Response Center within SEPA, and an Inter-Ministerial Committee for Environmental Protection under the State Council.[24]

If the evolving new crisis management system needed further testing, the year 2008 obliged. A first crisis was brought on by the unusual winter weather which affected 21 provinces, and caused 129 deaths, in the early part of the year. The snow and ice which afflicted South China had not been anticipated in contingency plans and led to a cascade of problems affecting the transportation and electric power systems, and producing widespread discomfort and economic loss. The emergency response system was slow to respond, again, demonstrating both the problems of information management and interagency coordination. In this case, the fact that the emergency cut across provincial boundaries further increased the administrative complexities of the crisis. The weather events also demonstrated the complex interdependencies that a modernizing China has developed, where a troubled energy policy, for instance, affected coal supplies, which in turn affected the ways in which crisis managers could, or could not, respond to transportation and heating emergencies.[25]

In response to the unusual snowfall which began to impact south and central China on January 10, the new Emergency Management Office, in an effort to promote an integrated interagency response, finally swung into action on January 22. As the weather worsened, unexpected problems, involving the jurisdictions of a number of different organizations, multiplied and overwhelmed the emergency response regime. By January 28, it was decided that a special disaster relief and emergency response command center had to be established under the

State Council. The new *ad hoc* command center, superseding the Emergency Management Office, then set up six task forces – for transportation of energy supplies, road clearing and repairs, powerline repairs, post-disaster recovery, and news media and public information.[26]

In their review of this case, He Guizhen, Lu Yonglong , and Zhang Lei concluded that the crisis was again exacerbated by recreancy. Vulnerabilities in the road maintenance system, in the electric power system (old transmission lines and an antiquated grid), and in the systems for coal deliveries were long neglected. Problems of interagency coordination, transportation planning, news reporting, and weather forecasting were made much worse by a lack of pre-event preparation. Response and recovery times, as a result, were unacceptably slow. Thus, what started as a freak act of nature, escalated into a major national crisis.[27]

An emergency of a somewhat different kind, with greater consequences for the loss of human life and property, was the Wenchuan earthquake in May, 2008. As China's most serious earthquake since the 1976 Tangshan event, the quake left some 69,000 dead and another 19,000 missing, and destroyed countless homes and facilities in western Sichuan. There is general agreement that China's crisis management system worked much better in this emergency than in the snow storm and in other prior emergencies. The response to the Sichuan earthquake was guided by the general Plan for Natural Disasters and by the National Emergency Plan for Earthquakes. Under the former, a Level I emergency was declared for the first time (the Yushu quake elicited a Level I response for the second time).[28] Response times were good, interagency coordination was relatively effective, information management was improved, and opportunities were available for civil society actors to participate in the response.

Yet, post-event analyses have nevertheless identified a number of problems needing further attention. In a study conducted for the Chinese Ministry of Civil Affairs and financed by the Asian Development Bank (ADB), for instance, Jiang Lingling, Wang Jiexiu, and Liu Lianyou identified a number of lessons learned from the Wenchuan earthquake.[29] They found that in spite of the prompt establishment of an earthquake disaster relief headquarters by the State Council to coordinate response activities, problems of coordination among central ministries, and between central ministries and local authorities still existed. A lack of attention to pre-event risk reduction clearly exacerbated the consequences of the disaster. An important part of the problem was inadequate preparedness and a lack of practice in emergency response. Since building collapses were the main cause of the severity of the consequences, the report argued that the quality of buildings – especially public buildings like schools, hospitals and centers for emergency management – must be improved by the introduction of strict building codes and by retrofitting buildings and homes.

The authors found that in spite of the new Emergency Management Law, China needs a stronger, more comprehensive disaster risk-management law which would spell out powers and responsibilities of various national level agencies and local authorities. The objectives would be to facilitate greater automaticity (and thus greater speed) in responding to events and require much more attention to risk reduction and mitigation. In addition, in spite of a much greater openness in the information environment relative to other disasters, the authors nevertheless call for more attention to the provision of prompt and accurate information for both the Chinese public and the international community, in order to "ensure transparency, facilitate operational effectiveness and achieve an appropriate level of interaction between the government and the public during a disaster."[30]

As widely reported, responses to the Wenchuan earthquake involved an outpouring of support from the public in ways that had not been seen before. This included generous donations of funds, active participation by non-governmental organizations (NGOs) in emergency response efforts, and the provision of psychological counseling services. The report found that more attention should be given to a "needs driven" approach to the management of donations, as opposed to a "supply driven" one, and to the provision of resources for the development of Community-Based Disaster Risk Management (CBDRM) processes to strengthen self-help capabilities. While the participation of NGOs was welcomed, the existing emergency response capabilities of NGOs was limited. There is therefore a need for more training and organization in the NGO community for disaster response. The latter observation also applies to the provision of psychological services. More professionalization and specialized training in responding to catastrophic incidents is needed.

Specialized professional rescue teams from both national and local agencies played important roles in responding to the earthquake. However, their numbers were inadequate with the result that needed assistance was slow to reach some of the affected areas. Since emergency rescue forces are based at dispersed locations around the country, it takes time for them to reach the site of a disaster, especially when the event is located in difficult terrain. Much more attention should be given, therefore, to the development of local rescue teams trained to be first responders.

The report found that China needed new mechanisms for involving those affected by the disaster in the processes of rebuilding their lives. It notes, for instance, that the supply of prefabricated dwellings did not effectively address the housing needs of local people, since they necessarily had to be located on flat land that would otherwise be used for cultivation. Through the assistance of a number of other provinces which have made commitments to the recovery process, reconstruction has proceeded quite rapidly. However, this has had the effect of driving up the prices of construction materials and labor to unreasonable levels and has led to building practices which do not necessarily incorporate earthquake-resistant standards. In spite of improvements in assessing damages and losses, China's practices in this area still fall below international standards. There is a need to introduce better methodologies for damage assessment and needs analysis, and to train specialists in their uses.

Although there is general agreement that the new Emergency Management Office acted far more credibly than in the snowstorm, its role was still subordinate to the *ad hoc* organization, the Earthquake Relief Headquarters established by the State Council to take command of the situation. According to the authors, "the country needs to establish a national disaster management (authorized) agency with sufficient power to target all hazards and disasters and implement a full range of disaster risk management strategies."[31] As discussed further below, the question of having an effective, permanent emergency response office has yet to be resolved in China.

"Learning from catastrophes"

As suggested at the outset, one can legitimately speak of contemporary China as part of today's "risk society."[32] The management of risks, and the crises that emanate from them, have clearly become a central part of governance for Chinese political leaders and for society as a whole. The failings of the crisis management regime in recent years have been studied and analyzed, and there is clearly an effort to "learn from catastrophe."[33] Reportedly, rescue efforts in the Yushu earthquake were more efficient and better organized as a result of lessons learned in Wenchuan,[34] and responses to the disasters of the summer of 2010, in general, were prompt and showed improved inter-agency coordination.

Many of the lessons learned are incorporated into the 2009 State Council White Paper, "China's Actions for Disaster Prevention and Reduction," which focuses on the steps taken to enhance disaster reduction capabilities during the eleventh Five-Year Plan – strategic goals, legal framework and institutional changes for disaster reduction.[35] It also addresses questions of public participation and international cooperation. In many ways, the White Paper reflects an awareness of the importance of comprehensive risk management noted at the outset.

In this process of learning, the lessons are not all from domestic emergencies. Instead, China has sought to learn from the national emergency management experiences of other countries and from participating in a number of international forums relating to disaster prevention, mitigation, and crisis management. As it does so, crisis management in China increasingly becomes professionalized, and the norms of international practice become points of reference for evaluating Chinese performance – consider, for instance, the involvement of the ADB in supporting an evaluation of the response to the Sichuan quake and the participation of the UN Environmental Program in reviewing the Songhua incident. The new professionalization, then, becomes the site for the dynamic interactions between international norms and those of China's own crisis management regime, interactions which show both the malleability, and lack thereof, of Chinese institutions. These dynamics are specially interesting in several important realms.

Civil–military relations

The Chinese armed forces have long been active in emergency response, often performing heroic deeds against very unfavorable odds.[36] Since 2000, efforts have been made to define more clearly the military's role in response to crisis incidents. In June, 2005, the State Council and Central Military Commission issued "Regulations on the Participation of the Military in Disaster Relief Operations," and in November, 2006, the CMC produced its "General Emergency Plan for the Military for Dealing with Sudden Incidents," a document which has not been made public.[37] As in other countries, when a disaster reaches a certain scale, only military organizations have the manpower, equipment, and command structure to respond to the needs. In spite of the positive reputation which the Chinese armed forces enjoy, problems with their roles have nevertheless been identified and, more broadly, the ways in which military and paramilitary personnel and organizations fit into a new modernized national crisis management scheme have yet to be defined.

Following the serious emergencies of 2008, the military initiated a review of its crisis management capabilities in the context of an overall review of "military operations other than war" (MOOTW), announcing in May, 2009 that five specialized emergency response units had been formed to deal with floods; earthquakes; chemical, biological, radiological, and nuclear incidents; major transportation accidents; and international disaster response and peacekeeping assignments.[38]

Reviews of the military's past performance by Chinese and foreign observers have pointed to problems in preparedness, including equipment problems and failures to develop specialized units for disaster response. For instance, although there were some units specialized in incident response, most military responders to the Sichuan earthquake had little rescue training. Deficiencies with the airlift capabilities of the PLA (People's Liberation Army) Air Force got particular attention. In general, the armed forces still lacked a broad range of engineering capabilities relevant to earthquake response, including a variety of necessary equipment such as heavy lift helicopters. Thus, in spite of rapid mobilization, the forces were not prepared to deal with rescues during the first critical 72 hours.[39]

There has been considerably less public discussion, however, of the ways in which military preparedness and responses to crises fit with the new institutions and policies on the civilian side. Although there is clearly evidence of civilian–military cooperation in responding to recent disasters, the maintenance of a military–civilian divide has been evident. In the Sichuan earthquake, for instance, military commanders reportedly resisted implementing orders from Premier Wen, who was then directing response efforts on the civilian side, pending clarification from higher authority, presumably, including the Central Military Commission.[40]

The command and responsibility relations for civil–military integration in the face of an emergency are seemingly quite ambiguous. China's 2007 Emergency Response Law and Regulations on the Participation of the People's Liberation Army in Emergency Rescue and Disaster Relief identify *both* the State Council and the Central Military Commission as command organizations in crises. In the Sichuan earthquake, the response was led by the special State Council Headquarters For Resisting Earthquake and Providing Disaster Relief, with Premier Wen Jiabao in command and vice-premiers Li Keqiang and Hui Liangyu as deputy commanders. The Headquarters established nine specialized working groups to respond to different aspects of the disaster with military or paramilitary organizations participating in six of these.[41] Yet, as has been customary, the PLA also established a separate "PLA Command Group for Resisting the Earthquake and Providing Disaster Relief." The Command Group was led by Chief of the General Staff Chen Bingde, with assistance from CMC Vice-Chairman Xu Caihou.[42]

In the Yushu earthquake, the reported response times were impressive. Word of the quake was received at General Staff Headquarters 12 minutes after the occurrence (indicating effective communication between the PLA and civilian earthquake authorities), most of the 12,000 military rescuers were at the site within 30 hours.[43] Other reports, though, indicated that the troops were not working on site as effectively as they were expected to, suggesting again a lack of preparedness. In spite of the rapid response, how the lines of command and responsibility between civilian and military authorities are worked out remains somewhat unclear. The idea of an integrated national crisis management regime is unlikely to be realized as long as the divide continues.

The role of non-state actors

As a number of reports have indicated, the participation of non-state actors in responding to the Wenchuan earthquake was somewhat unprecedented.[44] There was an outpouring of volunteerism of time and money, and a number of NGOs responded actively.[45] Press coverage from both the Chinese and international media was, at first, relatively open and quite extensive. In many ways, the earthquake seemed to mark a turning point in state–society relations in China, both pointing to an increasingly robust civil society, while also suggesting a distinctive relationship between the state and civil society groups.[46] Indeed, it seems as if the more effective NGOs were those that established partnership relationships with mass organizations such as the Communist Youth League, which mobilized large numbers of young people, and the Chinese Red Cross (a government organized NGO, or GONGO). Those who failed to establish these partnerships were more likely to be told to leave the area.

At the same time, there have been reports that non-state actors lacked preparation for effective emergency response and on occasion interfered with the efforts of government response teams, civilian and military. In the Yushu quake, the remoteness of the area made it difficult for non-state actors to reach the site. The one major exception, however, were Buddhist monks who, according to some reports, were very active in the rescue efforts. Reportedly, though, monks from outside of the immediate area who had traveled to Yushu were asked to leave, since they allegedly were interfering with the rescue and relief work, in spite of the fact that they had language and cultural identities with most of the victims, of relevance to relief efforts, which other rescuers lacked.[47] In addition, we should note the arrest of the Tibetan intellectual, Zhogs Dung (TraGyal), in spite of his reputation for being close to the Communist Party, for his efforts to organize relief work in response to the Yushu quake. Reportedly, he angered the regime by urging that private donations from Tibetans not be sent through the Chinese Red Cross, or official channels, but should instead be delivered directly in order to ensure that they reach those in need.[48]

It seems likely that non-state actors will, if anything, become more active in responding to future disasters, but in doing so they will be acting in a policy and institutional limbo, especially with regard to the three areas that most impeded their activities during the aftermath of the earthquake – registration with the state, organizational capacity, and fund raising.[49] As with the civil–military relations issue, it would appear that the development of an integrated crisis management system will require further clarification of the role of NGOs and other non-state actors, and opportunities for the latter to acquire the necessary crisis response capabilities so that they can be integrated more effectively with government efforts.

Risk reduction

There is little doubt that the challenges of crisis management in China would be more readily handled were risks reduced. While China has shown an impressive capacity to "learn from catastrophes" with regard to the enhancement of crisis response capabilities, it has seemingly been less successful at implementing lessons illustrating the importance of reducing risks. This point was made abundantly clear again during the summer of 2010 in the face of industrial accidents and the landslide in Zhouqu county, the catastrophic qualities of which were largely man-made. Thus, in spite of regular environmental emergencies, the prevention of environmental insults seems to be elusive – major environmental accidents during the first six months of 2010 were double those of 2009. In spite of the lessons learned about shoddy construction and urban design from the Tangshan and Wenchuan earthquakes, strict building codes have not always been enforced in spite of national retrofitting programs and policy declarations to the contrary.[50] One of the heartening aspects of the Wenchuan earthquake recovery efforts has been the use of a "buddy system" in which other provinces "adopt" communities in the affected area and pitch in with reconstruction assistance. This has made the recovery phase proceed quite rapidly, but a number of observers are concerned that building practices do not yet reflect the lessons of poor construction brought home so poignantly by the deaths and injuries of the 2008 quake.[51]

The insurance question

The frequency and severity of natural disasters in China in recent years has stimulated debate about the feasibility of, and possible roles for, catastrophe insurance. Large disasters tax the resources of the state – at the local and national levels – and point to the need for mechanisms to share risks and generate resources that could be applied to emergency response and recovery efforts. In addition, interest in catastrophe insurance is driven by the belief that insurance schemes reconfigure incentives and force attention to prevention and mitigation efforts. Thus far, disaster insurance is not widely used in China; only about

5 percent of losses are covered by insurance, in contrast to the 30 percent covered in developed countries.[52]

Nevertheless, some experimentation with catastrophe insurance has occurred. A flood insurance scheme was introduced in 1992 following the Huai River flood in 1991 but was subsequently abandoned, and programs for agricultural disaster insurance are being attempted. In spite of encouragement from international institutions like the World Bank, the challenges of establishing catastrophe insurance in China are daunting. Although some insurance companies are offering earthquake insurance, the penetration rate is low by international standards – only 3 percent in China in contrast to 50 percent in Japan and more than 20 percent in Turkey. While the Wenchuan earthquake raised awareness about the possible utility of earthquake insurance, the magnitude of the destruction was sobering to the companies, and they have subsequently delisted earthquake coverage from catastrophe policies. In March, 2009, the National People's Congress began discussions of a national catastrophe insurance law.[53]

Leadership training

As noted at the outset, one of the conundrums of crisis management is that the very nature of natural disasters, environmental emergencies, and serious industrial accidents is often one of surprise. Advanced planning and preparedness, of course, can make a big difference in effective responses, but there clearly are limits to planning when the contingencies of unexpected events confront us. For instance, established contingency plans were of limited value in the snowstorm emergency, as one problem after another combined to produce a fully unfamiliar and unanticipated crisis. Hence, crisis management systems must be nimble and adaptable.[54] The ability to respond to contingencies outside of established plans puts a premium on good leadership, and the cultivation of effective leadership should become part of the building of crisis management capabilities.[55] The recently established National Institute for Emergency Management (NIEM) at the Chinese Academy of Governance (formally the National School of Administration) indicates that China has begun to take crisis management leadership issues seriously. The ability to respond creatively to surprising contingencies is also being strengthened by the availability of new technologies – satellite observations, global positioning systems, modern telecommunications, ground sensors, etc. – which China is introducing or which are available from international partners.[56]

Information management

There is no doubt that there have been major changes in Chinese thinking about how information pertaining to emergencies and crisis situations should be managed. As the 2009 White Paper puts it, the release of information should be "prompt and precise, open and transparent." That said, it is also clear that Chinese media and information management practices still depart notably from the norms of international best practices. As noted above, media have been instructed to direct celebratory attention to the role of the military in disaster response and, more generally, there is still a strong inclination to manage information during crisis incidents in support of political objectives. In addition, there are also active efforts at information control during pre- and post-incident situations which, arguably, constrains knowledge of official recreancy and thus inhibits risk reduction efforts. The most egregious recent case of this information control is the draconian steps taken to suppress information about collapsed schools and the deaths of schoolchildren in Sichuan, as seen, for instance, in the imprisonment of Tan Zuoren and the beating of Ai Weiwei.

A national coordinating agency

It is clear that the Emergency Management Office set up in 2005 has yet to function as an authoritative crisis management agency.[57] Instead, in the face of Level I incidents, special *ad hoc* command centers under the State Council have been set up which presumably have been able to get some control over China's notorious "stovepiping" problems in the midst of a crisis. When the crisis passes, however, they are then disbanded, and the system reverts to a more fragmented set of agencies which have failed to give coordinated attention to pre-event risk mitigation and crisis preparation and to post-event recovery.[58]

Not surprisingly, therefore, inter-agency coordination seems to improve when crises are designated as serious, presumably because this leads to the involvement of high-ranking leaders, as seen, for instance, in Premier Wen Jiabao's association with the establishment of crisis command centers. Wen's actions indicate the growing political significance of recent crises, but some observers believe is counter-productive, since (1) the intervention of the Premier doesn't allow the provisions of the new crisis management regime to take hold and become institutionalized, and (2) the visit of such a high official to the site of a disaster requires that the attention of incident management leaders on the scene be shifted from the management of the emergency to the management of the visit.[59]

Some members of China's crisis management community believe that a stronger permanent central agency, perhaps inspired by the US FEMA, is called for. Others, however, are either unconvinced that such an office would work in the Chinese context, or are resistant to the idea for fear of losing power and resources over the crisis management turf they now control.

Vertical coordination and the tiao–kuai problem

Whereas much of the recent reform and innovation in crisis management institutions, discussed above, focused on horizontal coordination among government agencies, a variety of problems of vertical coordination among different levels of the state still plague the system. Although it sometimes functions as one, China is not a federal system in which sub-national powers and prerogatives are spelled out. Instead, as a unitary state – one with an enormous population and a large and complex landmass – it faces difficult problems of coordination between the central government and provincial, and sub-provincial governments. Local officials are part of a vertically organized government system of accountability (the *tiao*, or "branch"). At the same time, the Communist Party is expected to provide unified leadership across different levels of government; thus, local Party committees also oversee local government officials in a system of horizontal supervision (thus, creating the *kuai*, or "lump"), Party–government relations, in short, can and do further complicate the development of effective crisis management system.

Local governments, for instance, are intended to be first responders to emergency incidents, but local governments are also politically subordinate to local Party authorities who control the career prospects of local officials and the flows of information within the government structure. The "situational awareness" of local officials to emergency conditions is thus strongly influenced by the preferences of local Party officials, a fact which makes local decision-making and crisis-reporting an especially complex matter. Should a report of an incident be sent first to the central Party authorities or to the appropriate higher level of government? Should a report be sent at all and, if so, at what point in the development of the event? Should the report of an incident be understated (so as not to reflect badly on local authorities), or should it be reported as more serious (in order to "play it safe," should the incident take on crisis characteristics)? As noted above, the new crisis management regime relies heavily on the flow of information from the site of an incident, but the institutional complexities of center–local/Party–government relations often undercut the requirements for accurate and candid information.

Conclusion – the "Chinese model" and its limitations

A distinctive feature of China's approach to crisis situations is its remarkable ability to mobilize resources quickly at the time of a catastrophic event and allocate them to where they are needed. China's crisis management capabilities, thus, are impressive in comparison with those of many other countries; the response of the Government of Pakistan to the Indus River flood of the summer of 2010, for instance, stands in sharp contrast to China's recent record. But whereas rapid mobilization in the face of a crisis marks the strength of the "Chinese model," the maintenance of purpose and coordinated action needed for pre-event risk reduction, and post-event policy implementation to prevent a recurrence of the crisis, are marks of its weakness.[60] These strengths and weaknesses are manifestations of an inherited system of political institutions with which international best practices fit uneasily.

The reform of China's crisis management system, thus, is still a work in progress. The events of the last several years have tested the system, have stimulated learning, and have led to significant policy and institutional innovations.[61] At the same time, many problems remain and, being outgrowths of established institutions, will likely resist solution. But, as crisis incidents continue to multiply, as they surely will, the pressures for change will surely increase. China is thus in a race pitting a developing capacity for institutional innovation against the generation of risks from natural disasters (exacerbated by human interventions) and environmental and industrial safety emergencies. How the race is run will have a lot to do with the kind of society China will have in the coming years.

As China seeks to devise new organizations and procedures to harmonize Chinese realities with internationally recognized best practices, it has clearly made progress in preparing for crisis response, response implementation, and to a lesser extent, recovery. But, although there is a declarative commitment to the importance of pre-event prevention and mitigation, the prevalence of recreant behavior among officials and industrial leaders makes the promise of risk reduction somewhat empty to date. It is therefore difficult to escape the impression that the successful modernization of China's crisis management system ultimately depends on changing the incentives of critical decision-makers throughout the polity and economy such that risk reduction and risk mitigation investments prior to catastrophic incidents deserve much higher priorities. Whether this can be accomplished without fundamental changes in the information culture, and thus in political practice, remains to be seen.

Notes

1 In 2009, there were 80,000 casualties resulting from natural disasters. "Public Still Unprepared for Natural Disasters: Interview with Shan Chunchang," *Global Times*, April 25, 2010, at http://opinion.globaltimes.cn/commentary/2010-04/525687.html.

2 Information Office of the State Council, "China's Actions for Disaster Prevention and Reduction," *Renmin Ribao* [*People's Daily*], May 11, 2009.

3 Richard P. Suttmeier, "The 'Sixth Modernization'? China, Safety, and the Management of Risks," *Asia Policy*, No. 6 (July, 2008); Paul Thiers, "Risk Society Comes to China: SARS, Transparency and Public Accountability," *Asian Perspectives*, Vol. 27, No. 2 (2003), pp. 241–251.

4 These include the Hongdong quake of 1303 in Shanxi, the Huaxian earthquake of 1556 in Shaanxi (thought to be the world's most deadly), the 1920 Haiyuan quake in Ningxia, and the Tangshan earthquake of 1976. Historical records also indicate that there had been eight magnitude-7 earthquakes within 200 km of the 2008 Wenchuan epicenter. Zifa Wang, Tun Lin, and George Walker, "Earthquake Risk and Earthquake Catastrophe Insurance for the People's Republic of China," MPRA (Munich Personal RePEc Archive), report prepared for the Asian Development Bank, MPRA Paper No. 21248 (March 10, 2010), p. 1, posted at http://mpra.ub.uni-muenchen.de/21248/.

5 Robert J. Antony and Jane Kate Leonard (eds), *Dragons, Tigers, and Dogs: Qing Crisis Management and the Boundaries of State Power in Late Imperial China* (Ithaca, NY, East Asia Program: Cornell University, 2002).

6 Recreancy is understood here as "the failure of institutional actors to carry out their responsibilities with the degree of vigor necessary to merit the societal trust they enjoy." Recreancy is thus a failure of individuals or institutions to live up to expectations of "trust, agency, responsibility, or fiduciary . . . obligations." William R. Freudenburg, "Risk and Recreancy: Weber, the Division of Labor, and the Rationality of Risk Perceptions," *Social Forces*, Vol. 71, No. 4 (1993), pp. 909–932.

7 Judith Shapiro. *Mao's War against Nature* (Cambridge and New York: Cambridge University Press, 2001).

8 For instance, at the time of the 1976 Tangshan earthquake, one of the most severe natural disasters in the pre-reform period, there was no specialized disaster management agency and, although a crisis management center was promptly established, there was no significant planning or preparedness, or established protocols, for coordinating various agencies. Although 100,000 troops were mobilized for the rescue effort, there were no professional rescue teams or specialized rescue equipment to speak of. It has been estimated that of those rescued, some 80 percent were saved by local people themselves. In addition, the government refused offers of assistance from other governments and did not have the benefits of participating in the international regimes for disaster risk reduction and emergency response. Poor building construction accounted for most of the deaths, but inadequate attention was given to building standards and land-use planning in the reconstruction efforts. Jiang Lingling, Wang Jiexiu, and Liu Lianyou, "People's Republic of China: Providing Emergency Response to Sichuan Earthquake," Asian Development Bank, Technical Assistance Consultants Report, Project No. 42150 (December, 2008), pp. 132–133.

9 Kirk Smith, "The Risk Transition," *International Environmental Affairs*, Vol. 2, No. 3 (1990), pp. 227–251.

10 Aaron Wildavsky, *Searching for Safety* (New Brunswick, NJ: Transaction Press, 1988).

11 China's first disaster reduction plan was developed in conjunction with the United Nations Development Program in 1998.

12 Herman B. Leonard and Arnold M. Howitt, "Acting in Time Against Disasters: A Comprehensive Risk-Management Framework," in Howard Kunreuther and Michael Useem (eds), *Learning from Catastrophes: Strategies for Reaction and Response* (Philadelphia: Wharton School Publishing, 2010), pp. 18–40.

13 Arjen Boin, Paul Hart, Eric Stern, and Bengt Sundelius, *The Politics of Crisis Management: Public Leadership Under Pressure* (Cambridge and New York: Cambridge University Press, 2005), pp. 2–4.

14 Arnold M. Howitt and Herman B. Leonard, *Managing Crises* (Washington, DC: CQ Press, 2009), pp. 1–14, 611–624.

15 "Tragedy of Errors in Gas Blowout," *China Daily*, January 12, 2004.

16 As Xue and Zhong point out, for familiar threats from flooding and drought, earthquakes, typhoons, etc., the government's lead agencies had developed contingency plans which could be promptly employed. The threat from SARS, however, being unfamiliar, revealed limitations to the "cognitive capabilities" of the crisis management regime. Lan Xue and Kaibin Zhong, "Turning Danger to Opportunities: Reconstructing China's National System for Emergency Management After 2003," in Kunreuther and Useem (eds), *Learning from Catastrophes*, pp. 190–210.

17 Drew Thompson and Carla Freeman, *Flood Across the Border: China's Disaster Relief Operations and Potential Response to a North Korean Refugee Crisis* (Washington, DC: US–Korea Institute, Johns Hopkins University School of Advanced International Studies, 2009), pp. 37–38.

18 Xue and Zhong, "Turning Danger to Opportunities," pp. 190–210. Some cities have invested heavily to produce a high-tech emergency response system. Nanning, for instance invested 170 million RMB for a system inspired by the US 911 system which involves technical cooperation with Motorola. Lei Zhang and Lijin Zhong, "Integrating and Prioritizing Environmental Risks in China's Risk-Management Discourse," *Journal of Contemporary China*, Vol. 19, No. 63 (January, 2010), p. 132.

19 James Bellacqua, *A Closer Look at Chinese Emergency Planning*, CNA (April, 2007). Cited with permission of the author.

20 James Bellacqua, *A Closer Look at Chinese Emergency Planning*.

21 The World Bank, *Water Pollution Emergencies in China: Prevention and Response* (Washington, DC: The World Bank, 2007). See, also, Nat Green, "Positive Spillover? Impact of the Songhua River Benzene Incident on China's Environmental Policy," China Environmental Health Project Research Brief, China Environment Forum (March, 2009), available at http://www.wilsoncenter.org/index.cfm?topic_id=1421&categoryid=EF5B2D71-423B-763D-DB0B54DDAE395532&fuseaction=topics.doc_topics&doc_id=514852&group_id=233293. The Songhua incident has been widely reported and analyzed, in large part because of the international implications of the spill. However, chemical spills occur regularly in China without getting the public attention that Songhua did (Interview, MEP official, March, 2010). Indeed, in the 2½ months following the Songhua spill, SEPA received 45 reports of significant environmental incidents, including the release of cadmium into the Beijiang River in Guangdong. Zhang and Zhong, "Integrating and Prioritizing Environmental Risks in China's Risk-Management Discourse," p. 124.

22 Zhong Kaibin, "Crisis Management in China," *China Security* (Winter, 2007), p. 93.

23 United Nations Environmental Program, *The Songhua River Spill, December 2005: Field Mission Report* (January, 2006), available at http://www.unep.org/PDF/China_Songhua_River_Spill_draft_7_301205.pdf.

24 The World Bank, *Water Pollution Emergencies in China: Prevention and Response.*

25 As one Guizhou-born interlocutor in Beijing explained to me, snow and severe winter weather were not unheard of when he was growing up in Guizhou. But at that time, communities were not linked together by modern transport systems, and by economic interdependencies, with the result that the sense of crisis was muted, if it existed at all. Personal communication (Beijing, March, 2010).

26 The snowstorm case illustrated how the incident classification system can change as the crisis deepens. Initially, the problems did not call for a national response. As the crisis worsened, a Level IV emergency was declared. This then led to a further escalation to Level II, and finally to the special arrangements involving State Council control. James Bellacqua, *Crisis Response in Action: Examining the Implementation of China's Emergency Response Plans*, CNA (October 10, 2008). Cited with permission of the author.

27 Guizhen He, Yonglong Lu, and Lei Zhang, "Risk Management: Lessons Learned from the Snow Crisis in China," *China Environment Series*, Issue 10 (2008/2009), pp. 143–149. See, also, Bellacqua, *Crisis Response in Action.*

28 Bellacqua, *Crisis Response in Action.*

29 Jiang, Wang, and Liu, "People's Republic of China," pp. 127–133.

30 Ibid., p. 128.

31 Ibid., p. 127.

32 Ulrich Beck, *Risk Society: Toward a New Modernity* (London: Sage, 1992).

33 Howard Kunreuther and Michael Useem (eds), *Learning from Catastrophes: Strategies for Reaction and Response* (Philadelphia: Wharton School Publishing, 2010).

34 "Public Still Unprepared for Natural Disasters."

35 "Full text: China's Actions for Disaster Prevention and Reduction," *People's Daily Online*, May 11, 2009, available at http://english.peopledaily.com.cn/90001/90776/90785/6655019.html.

36 For instance, since 1998, PLA engineering units have been assigned flood response responsibilities near major rivers. See, Peter J. Brown, "China Plans for the Next Big Disaster," *Asian Times Online*, May 30, 2009, at http://www.atimes.com/atimes/China/KE30Ad01.html. Reportedly, over 130,000 PLA and PAP (People's Armed Police) troops from all seven military regions were mobilized to respond to the Wenchuan earthquake and began to move supplies to the region on the first day. See also, Harold M. Tanner, "The People's Liberation Army and China's Internal Security

Challenges," in Roy Kamphausen, David Lai, and Andrew Scobell (eds), *The PLA at Home and Abroad: Assessing the Operational Capabilities of China's Military* (Seattle, WA: National Bureau of Asian Research, 2010), pp. 237–293. While the courage and heroism of many members of the armed forces cannot be doubted, it is also important to note the propaganda objectives of much reporting about the roles of the PLA and PAP in responding to emergencies. In the Sichuan earthquake, there were reports that media attention was directed away from civilian responders and NGOs to the military, and similar reports have surfaced in conjunction with the recent Qinghai quake, where media attention focused on the military rather than the activities of Buddhist monks and other civilian responders who were very active in the rescue. In addition, much of the reporting of the military's role in responding to the Sichuan earthquake called attention to the military's loyalty to the Party and the directives of Hu Jintao. See, James Mulvenon, "The Chinese Military's Earthquake Response Leadership Team," *China Leadership Monitor*, No. 25.

37 Thompson and Freeman, *Flood Across the Border*.

38 Brown, "China Plans for the Next Big Disaster." See also, Tanner, "The People's Liberation Army and China's Internal Security Challenges."

39 Jake Hooker, "Quake Revealed Deficiencies in China's Military," *New York Times*, July 2, 2008. See also Tanner, "The People's Liberation Army and China's Internal Security Challenges."

40 Interview in Beijing in March, 2010.

41 Bellacqua, *Crisis Response in Action*. Eight working groups were established at first on May 12, with a ninth added on May 15.

42 Mulvenon, "The Chinese Military's Earthquake Response Leadership Team."

43 "Chinese Army Makes Quick Response to Yushu Quake," *Xinhua*, April 20, 2010 at http://www.china.org.cn/china/qinghai_earthquake/2010-04/20/content_19868480.htm.

44 NGOs began to make an appearance in emergency response situations in the late 1990s (e.g., for flooding on the Yangtze), and in the SARS epidemic and in the earlier snowstorm. The scale of participation for the earthquake, however, far exceeded those previous experiences. One estimate put the number of NGOs (of all sorts) active in the earthquake region one week after the event at 50 (30 from Sichuan). Shawn Shieh and Guosheng Deng, "An Emerging Civil Society: The Impact of the Sichuan Earthquake on Grassroots Associations in China," Unpublished paper (December, 2009). Cited with permission of the authors.

45 Shieh and Deng report that more than 30 billion yuan in public donations poured in following the earthquake, an impressive figure when compared with the more than 16 billion spent from the government's earthquake fund two weeks after the event. See Shieh and Deng, "An Emerging Civil Society."

46 Jessica C. Teets, "Post-Earthquake Relief and Reconstruction Efforts: The Emergence of Civil Society in China?" *The China Quarterly*, No. 198 (June, 2009), pp. 330-347.

47 http://loudcanary.com/2010/04/28/the-stage-the-yushu-earthquake-in-fact-and-fiction/.

48 Jane Macartney, "Tibetan Writer Zhogs Dung Held for Organizing Yushu Quake Donations," *The Times Online*, April 26, 2010 at http://www.timesonline.co.uk/tol/news/world/asia/article7108169.ece.

49 Shieh and Deng, "An Emerging Civil Society."

50 The White Paper, for instance, calls attention to the need for stricter building codes and for retrofitting schools and residences to meet stricter codes. See, also, Cara Anna, "China Quake Awakens New Fears for School Safety," at http://www.ajc.com/news/nation-world/china-quake-awakens-new-475486.html.

51 Interview in March, 2010.

52 "Catastrophe Insurance in the Works," *People's Daily Online*, April 20, 2010 at http://english.people.com.cn/90001/90778/90859/6956566.html (accessed on April 21, 2010).

53 Wang, Lin, and Walker, "Earthquake Risk and Earthquake Catastrophe Insurance for the People's Republic of China." See, also, George Walker, Tun Lin, and Yoshiaki Kobayashi, "Is Flood Insurance Feasible? Experiences from the People's Republic of China," ADB Sustainable Development Working Paper Series No. 5, Asian Development Bank (April, 2009); "Catastrophe Insurance Policy for China," a report prepared for the World Bank Good Practice Notes (July 2008). Wang, Lin, and Walker, "Earthquake Risk and Earthquake Catastrophe Insurance for the People's Republic of China," p. 1.

54 Plans for the snowstorm emergency called for the closing of major roads until they could be cleared. Local authorities, however, contravene the plans and opened roads with strictly reduced speed limits. This had the effect of overcoming some of the more serious transportation bottlenecks while also helping to wear down snow and ice on the road surface.

55 On this point, I have profited from discussions with Professor Xue Lan of Tsinghua University.

56 See, for instance, Wang Guilian, Lu Yonglong, and Xu Jian, "Application of GIS Technology and Chemical Emergency Response," *Journal of Environmental Sciences*, Vol. 12, No. 2 (2000), pp. 172–177.

57 As noted by Shan Chunchang, head of the State Council expert group on emergency response, the Yushu quake experience indicated both that progress had been made, but that still more is necessary:

> First, we need to enhance the coordination between departments and all levels of government to better allocate the rescue resources. Due to the lack of coordination, our earthquake rescue teams, firefighters and medical services fail to form a link service. Second, the organizing of volunteers should be more orderly.

See "Public Still Unprepared for Natural Disasters."

58 Zhong Kaibin, "Zhongguo shi yingji: cong pingzai zhuanhuan dao pingzaijiehe" (The Chinese Emergency Response System: Transforming the Ordinary to the Coordinated) in *Zhongguo gaige* (China Reform) (forthcoming in 2011).

59 Of course, the presence of a national leader at the scene of a crisis, by signaling national care and responsibility, is an important political gesture. The absence of such a presence during Indus River flood in Pakistan stands as a notable contrast to Chinese practice.

60 Cf. Zhong, "Zhongguo shi yingji."

61 This includes the development of an international disaster response team which is being expanded from 220 members to 500, along with a significant increase in the budget for international and domestic rescue teams. The latter includes some 5,000 members distributed around the country. Tania Branigan, "China Boosts International Rescue Squad to Match Its Growing World Role," *The Guardian*, February 26, 2010.

7 China's management of natural disasters

Organizations and norms

Gang Chen

China is a continental state vulnerable to frequent natural disasters, which today affect about 300 million people every year. With its vast territory and diverse climatic and geological conditions, the country is exposed to most types of natural hazards like meteorological disasters, earthquakes, mudslides, marine disasters, biological disasters, and forest fires. Studies have shown that both in ancient history and modern days, China has been suffering the most from floods, earthquakes, and other meteorological disasters like drought and typhoon.[1] In response to devastating natural disasters that may pose fatal threats towards their governance, successive dynasties in ancient Chinese history developed complicated bureaucratic systems to prevent and handle those calamities.

Natural disaster management had become such a life or death issue for each dynasty that neglect of such duties on the part of the emperor would cause famine, riots, peasant rebellion, and finally, the demise of the imperial regime. Handling natural disasters like floods and droughts has become very much a part of China's rural life and in fact, part of China's history.[2] China's oldest dynasty, the Xia, was founded by Da Yu, a hero who successfully regulated the flood and taught the people how to tame China's rivers and lakes.[3] In the modern era, since the Chinese Communist Party (CCP) seized power from the short-lived Kuomintang regime and established the People's Republic of China (PRC) in 1949, the CCP-led government has been trying to improve the effectiveness of such disaster management through modernizing relevant bureaucratic organizations and establishing standardized norms and guidelines to direct such work that is vital for the survival of the communist regime.

Beginning with a brief review of China's natural disaster management before the start of reform and opening-up in 1979, the chapter mainly focuses on the PRC's modernized bureaucratic organizations in charge of natural disasters and the evolution of their operation mode in the past three decades. Based on analysis of China's thematic disaster response plans and national disaster management mechanisms, it discusses what kind of specific norms and guidelines have been incorporated into the interrelated stages of mitigation, preparedness, response, and recovery in the whole disaster management cycle, and to what extent they have been effective in limiting the devastating impact of natural calamities.

History of China's natural disaster management before 1979

Famine politics (huangzheng) *in ancient China*

China's recorded history has been full of devastating natural disasters like floods, droughts, earthquakes and pestilence. In the 5,000 year-long history, the Chinese people's fight against these natural calamities has never stopped, with major imperial regimes being able to formalize systematic institutions to manage natural disasters. Historians have called such disaster and famine-relief governance in ancient China famine politics (*huangzheng*),[4] which included policies, practices, institutions, and even theories related to the preparedness, relief, and recovery in the disaster management cycle. Most dynasties regarded such disaster-relief governance as an unshirkable and vital public responsibility for both the emperor and the officials, who were required to be directly involved in disaster and famine relief work under strict regulations and laws. As early as in the Western Zhou Dynasty (1046–771 BC), some officials in the central government started to be given disaster relief responsibilities like distribution of aid materials to disaster-stricken populations.[5] During China's Sui (AD 581–618), Tang (618–907), Song (960–1279), and Ming (1368–1644) Dynasties, the Ministry of *Hu* (*hu bu*), an analogy to today's ministry of finance and civil affairs, was appointed to take charge of the disaster relief work, together with other issues like taxation and household registration. Besides the Ministry of *Hu*, officials in other ministries, local governors, and even the emperor were still often involved in the famine politics (*huangzheng*). In the *Book on Disaster Relief and Lifesaving (Jiuhuang huomin shu),* an influential disaster-relief book written by Dong Wei in the Song Dynasty, the author said clearly that in the imperial hierarchy from the emperor, prime minister, province governor down to the county magistrate, different posts had different responsibilities to manage natural disasters.[6] According to Dong Wei, the emperor needed to practice self-reflection, have a thrifty lifestyle, improve governance and send ministers to distribute reserves in the granaries, while local officials were required to investigate and report the disaster consequences as well as dispense disaster relief materials and maintain social order. In the final imperial dynasty of Qing (AD 1644–1911), the central government had established a regular system of sending officials to disaster-hit regions for overseeing or assisting local relief work. Yet despite the intensive attention paid to the famine politics (*huangzheng*), China's imperial dynasties failed to establish an independent department specialized at disaster management.

Ancient dynasties in China had formalized complicated guidelines and institutions related to detailed procedures and measures in anticipation of and in response to natural disasters. Ancient Chinese regimes managed to institutionalize the build-up of water conservancy projects and national grain reserve as preventive and mitigation measures against frequent floods, droughts, and other natural disasters. Before Qin united the whole of China in 221 BC, warring states had built small-scale water management bureaucracies to regulate floods and droughts.[7] About 2,000 years ago, during the Han Dynasty (206 BC–AD 220), a special post of Director of Water Conservancy (*tu shui*) was first created by the government to formalize water management and floods control work. The Han Dynasty also had full-time officers for river management and established river management units at city and county levels. Various kinds of assistance like grain transfer and distribution, monetary grant, medical aid, control of crop price, loan, tax reduction, and migration had been recorded in Chinese history as important means to relieve people in disasters. As early as in the Western Zhou Dynasty, special posts were set up to manage grain reserves in the granaries in preparation for disasters and famines.[8] Since then, almost all the major dynasties along ancient history improved their build-up of grain reserves as an important preparatory measure against disasters and famines. Facilitated by the nationwide establishment of granaries, the ancient regimes very frequently granted grain and other kinds of foods to the disaster-hit population for the alleviation purposes. History has recorded numerous cases of government-sponsored porridge distribution to hungered people after severe natural disasters like floods and droughts. In the early Ming Dynasty, when Emperor Zhu Yuanzhang realized the time delay of one or two months caused by the complicated bureaucratic process of approving grain grants may cause more deaths, he issued documents authorizing local officials to distribute grain to the disaster-hit population before they got the formal permission from the capital.

Besides grain and food, monetary and medical aid was also frequently provided by China's ancient regimes for disaster alleviation. As early as in the Han Dynasty, it was recorded in the history that each family with casualties in the floods or earthquakes could get money compensation from the government. From the Han Dynasty to the Qing Dynasty, history had recorded occasional distribution of medicine and medical services to disaster-stricken areas. In the recovery stage of disaster management, ancient dynasties sometimes intervened in the market through changing the grain supply and demand to stabilize its price. According to the practice called rice trading (*pingtiao*), the government would sell more to the market from its national grain reserve in the disastrous year and increase its purchase in the bumper year. After the disasters, previous dynasties from time to time exempted or reduced taxes upon the people and loaned seeds or money to them for the quick economic recovery.

Although most ancient Chinese dynasties had not established special government departments to manage natural disasters, the emperors and their bureaucrats paid close attention to the preparatory and alleviation measures and therefore formalized a series of policies, practices, institutions, and even theories in this regard. Besides detailed practices like the establishment of national granaries, improvement of water conservancy projects, grain, monetary and medical aid, and rice trading, relevant institutions relating to disaster reporting as well as rewards and punishment of accountable officials had also been set up to ensure the effectiveness and efficiency of disaster relief. Nevertheless, due to the science and technology constraint and the lack of modern management institutions and norms that demand systematic integration of the various stages and relevant departments involved in the disaster management cycle, natural disasters in many cases still caused severe casualties and huge economic losses. Ancient governments paid more attention to short-term relief measures after the disasters rather than time-consuming preparatory and preventive actions. Such imbalance contributed to the recurrence of deadly flooding along the major rivers and lakes in China's history.

The Republic of China period (1912–1949)

After the Republic of China (ROC) took power from the monarch of the Qing Dynasty and set up its government in 1912, modernized institutions managing natural disasters were introduced with special departments appointed to take charge of relevant affairs. In August 1912, the Ministry of Internal Affairs (*nei wu bu*) was set up inside the ROC Beijing government to take charge of disaster relief work, philanthropy, and public health, with its subordinate department of civil affairs taking over the detailed disaster alleviation work.[9] Besides this full-time organization managing natural disasters, the ROC Beijing government also established an *ad hoc* national committee on disaster relief after the outbreak of catastrophes to coordinate the responses from the Ministry of Internal Affairs, the Ministry of Agriculture, the Ministry of Commerce, and Ministry of Transportation. In May 1921, the Ministry of Internal Affairs publicized a statute to institutionalize a national committee on disaster management, which aimed to find out the causal factors behind natural disasters and work out corresponding preparation and prevention plans.[10] This national committee had six divisions respectively in charge of general affairs, farmland and forestation, water conservancy, grain reserve and distribution, migration, and protection of labor forces. As early modern effort to comprehensively manage the Yellow River, known both as "Cradle of Chinese civilization" and "China's sorrow," the Chih-li River Commission, and later the Hua Bei River Commission began the first systematic monitoring of the Yellow River flow in 1922.[11]

After the Nationalist Party (*Kuomintang*, KMT) established a government with Nanjing as its capital and controlled most of China in 1927, they basically copied the previous disaster management model set up by the Beijing government with the Ministry of Internal Affairs still being the full-time apparatus on this issue. In response to the Central China Floods in 1931, probably the deadliest natural disaster ever recorded, the ROC Nanjing government established a national committee on flood relief to coordinate the work done by different government departments. In the recovery stage, the Nanjing government promoted the "Huai River Conservancy Commission" to address flood problems. The Yellow River Water Conservancy Commission was founded in 1933 to ease China's historically problematic inter-provincial tensions upon the management of the indocile river. However due to the Anti-Japanese war, the Chinese Civil War and the lack of funding, the KMT government made little progress in building up effective water conservancy infrastructures along the major rivers. To achieve political and military goals in the early stage of the Anti-Japanese War, the KMT government even opened up the dike in Huayuankou without informing the public beforehand and created the 1938 Yellow River flood in an attempt to halt the rapid advance of the Japanese invaders. As a result of this man-made natural disaster, the course of the Yellow River was diverted southwards for nine years afterward, inundating 54,000 km^2 of land in central China and took at least half a million lives. Thanks to the reestablishment of the dikes in 1946 and 1947, the Yellow River returned to its pre-1938 course after the end of the Anti-Japanese war.

During the ROC period, China was in the early stages of industrialization and modernization that facilitated the nation to introduce modern-style bureaucracy to manage natural disasters. Yet since the ROC's nation-building process was often disrupted by large-scale wars and severe political confrontation, the distracted bureaucratic system failed to function properly to address major natural disasters. Despite the KMT's high-profile plans to improve water conservancy works and fundamentally solve the problem of floods along major rivers like the Yellow River and the Huai River, the regime's warfare priority had from time to time prevented it from being a responsible government in disaster mitigation and post-disaster recovery.

PRC pre-reform period (1949–1978)

Since the CCP defeated the KMT in a civil war and founded the PRC in 1949, the new regime in its early days continued to authorize the Ministry of Internal Affairs (*nei wu bu*) to be responsible for disaster relief work. In February 1950, a central committee on disaster relief (*zhongyang jiuzai weiyuanhui*) was established to coordinate disaster management work among the Ministry of Internal Affairs, the Ministry of Finance, the Ministry of Agriculture, the Ministry of Water Resources, the Ministry of Health, and other related ministries and governmental departments, with Vice Premier Dong Biwu being the head of the committee. Following the central government's instruction, local governments subsequently set up their disaster relief committees at the prefecture, county, and township level. During the frenetic "Great Leap Forward" movement in 1958, the central committee on disaster relief and many of its local branches were abolished amidst people's blind faith that natural disasters would no longer pose serious threat in communistic China. In 1969, two years after the start of the Cultural Revolution, the Ministry of Internal Affairs was abolished with most of its disaster management functions scattered in other ministries and departments, which led to the inefficiency and even chaos in this area.

In the conference marking the establishment of the central committee on disaster relief in 1950, Dong Biwu put thrift, self-dependence, mutual aid, and using labor instead of aid materials as guidelines for new China's disaster relief work.[12] When the young PRC joined the socialist camp in the early stage of the Cold War, it enunciated that it refused to accept any foreign aid offered by the antagonist western camp, especially the United States, for the disaster relief in China. Since the new regime was not well equipped to provide necessary monetary and material aid, the PRC government at that time took a lot of effort in asking disaster-hit population to lead thrifty lives and resume production as soon as possible.

Influenced by the Cold War mentality, the PRC in its first thirty years took very cautious attitudes towards foreign aid, refusing to accept most of them even in severe disasters like the 1976 Tangshan Earthquake. At the time of the magnitude 7.8 quake that hit Tangshan at the end of the Cultural Revolution (1966–76) and took at least 240,000 lives, the insular PRC leadership clung to extreme-leftist ideology and refused all offers of international aid, even from such neutral bodies as the United Nations and the Red Cross. Instead, the Chinese government urged its citizens to "Resist the Earthquake and Rescue Ourselves."[13] Such self-reliance attitudes toward disaster relief effort reflected the young PRC's oversensitivity over the sovereignty issue based on memory of China's tribulations caused by foreign invasion and interventions in the past a hundred years. In less than one year after the PRC was founded, the United States expressed its concern over the ongoing famine in China and its willingness to offer help, but the Communist leadership interpreted such a posture as imperialists' intention to undertake subversive activities. In 1954, China briefed foreign journalists on the reasons behind the self-reliance principle, saying that

> since China is a unified country with vast territory, diversified resources and a large population, it is able to use its own domestic resources to handle natural disasters that hit certain regions every year when other parts of China are enjoying a bumper year.[14]

The PRC's disaster management work had been dominated by the purely self-dependence guidelines until 1979 when the Communist Party decided to start reform and an open-door policy.

The nascent regime had bitter experiences in managing disasters in the first decade of the CCP ruling when the country witnessed two major floods in the Yangtze River Basin in 1954 and 1956, as well as the beginning of the three-year famine (1959–61) that was officially blamed on natural disasters. The CCP used "The Three Years of Natural Disasters" (*sannian ziran zaihai*) to describe the famine in which at least 20 million people died of starvation caused by the "Great Leap Forward" and other utopian government policies. Since then, the PRC government has drawn lessons from such miserable experiences and made remarkable progress in disaster preparation and mitigation. After the 1954 Yangtze River floods that occurred mostly in central Hubei Province and killed about 33,000 people, the building of hydroelectric dams and reservoirs as a solution to flooding and energy shortage gradually became commonplace, and millions of peasants during Mao's time were mobilized to construct small and medium water conservation work.[15] Such efforts over the years had the cumulative effect of alleviating floods through the 1960s and 1970s. In his poem "Swimming" (1956), Mao Zedong envisioned future dam construction on the Yangtze as "to hold back Wushan's clouds and rain till a smooth lake rises in the narrow gorges; the mountain goddess if she is still there, will marvel at a world so changed.''

Since there were no major earthquakes in the first 17 years of the CCP ruling, the PRC had not paid much attention to earthquake management until the deadly quakes struck Xingtai, Hebei province in 1966 and took about eight thousand lives. China set up the National Earthquake Administration to take charge of earthquake monitoring, research, and emergency responses in 1971. In 1975, China successfully predicted strong quakes in densely populated Haicheng, Liaoning Province, and thus saved more than a hundred-thousand people. In 1976, despite its failure in predicting the Tangshan Earthquake, China still managed to forecast devastating earthquakes respectively in Sichuan and Yunnan Province.

Natural disaster management bureaucracy in the reform era

Reform-related bureaucratic structure

Since the beginning of the reform and opening up, the PRC has gradually normalized the bureaucratic system in charge of natural disaster management once disrupted by the anarchical Cultural Revolution. In 1978, two years after the Cultural Revolution was over, the Ministry of Civil Affairs (*min zheng bu*) was established to resume part of disaster management functions. During most of the 1980s, the Ministry of Civil Affairs was only in charge of disaster relief in rural areas, while the State Economic Commission (SEC) owned the power to coordinate all the disaster management work at the national level. In 1988 when the SEC was abolished in another round of administrative reform, such responsibility was shifted to the State Planning Commission, which later transferred this function to the State Economic and Trade Commission established in 1993. In 1998, according to a circular issued by the State Council General Office, the State Economic and Trade Commission officially handed over the function of organizing and coordinating disaster relief work to the Ministry of Civil Affairs,[16] which since then institutionalized its pivotal role in managing natural disasters in China.

In 1989, in response to the UN resolution to designate the 1990s as the International Decade for Natural Disaster Reduction (IDNDR), the Chinese government founded the National Commission for the International Decade on Natural Disaster Reduction, which comprised more than 30 ministries and commissions, including the Ministry of Civil Affairs, the Foreign Ministry, the State Planning Commission and relevant military agencies and social groups. It soon became the top-level decision-making panel that commanded and coordinated the disaster relief work scattered in different ministries and departments. In 2000, as soon as the IDNDR was over, it was renamed the China Commission for International Disaster Reduction, which turned out to be a regular inter-agency coordination body under the State Council rather than an interim organization in answer to the call from the United Nations. In January 2005, it was renamed the China National Committee for Disaster Reduction (NCDR), headed by a Vice Premier of the State Council, and was responsible for studying and formulating principles, policies and plans for disaster reduction, coordinating major disaster activities, giving guidance to local governments in their disaster reduction work, and promoting international exchanges and cooperation.[17] Since then, the Chinese government has formalized the NCDR as the top inter-agency mechanism in charge of disaster management, with the Ministry of Civil Affairs taking on routine jobs of the NCDR and playing a pivotal role in coordinating work done by various ministries and sectors.

According to the division of work under current disaster management bureaucracy, the Ministry of Civil Affairs, which deals with most of the daily administrative work of the NCDR, coordinates the disaster relief work among various departments, collects and releases the latest disaster situation information, manages and distributes the aid fund and materials from the central government, cooperates with other departments in urgently allocating aid to disaster-hit people, prepares the nation-level disaster relief material reservation, organizes public donation and undertakes international cooperation in this area.[18] The Ministry of Finance formulates the yearly budget fund for national disaster relief, supervises the use of such funds and provides special funding for severe disasters, while the State Administration of Taxation formulates and implements the tax-waiver policy in disaster-hit regions. The Ministry of Agriculture is charge of the provision of agricultural materials like fertilizers and diesels in the disaster-hit regions and helps peasants to produce grain in the recovery phase. The Ministry of Water Resources plays a vital role in fighting against major floods and droughts and thus undertakes the administrative work of the State Flood Control and Drought Relief Headquarters, another inter-agency mechanism to cope with severe floods and droughts. The Ministry of Health is in charge of the medical services in the disaster-hit regions, while the Ministry of Transport repairs the roads and railways damaged in natural disasters. The China Earthquake Administration forecasts and monitors the earthquakes, while the China Meteorological Administration takes charge of weather forecasts and releases warnings on meteorological disasters.

In 2002, the National Disaster Reduction Center (NDRC) was set up under the Ministry of Civil Affairs to serve as an information sharing platform on various natural disasters. It provides policy consultancy services and technical support for ministries and sectors involved in disaster management, collects and analyzes disaster information, and trains professions in the disaster reduction area.[19]

Ad hoc *inter-agency mechanisms*

Besides the China National Committee for Disaster Reduction as the regular inter-agency mechanism on comprehensive disaster management, the Chinese government has established other *ad hoc* leading panels to direct and coordinate relief work among various ministries and sectors when the country is hit by severe floods, droughts, earthquakes, and forest fires.

Of all types of natural disasters, frequent meteorological disasters like floods, droughts, and storms received the most attention from the Chinese people. China's major river basins are well within the monsoon zone of the Pacific, with over 50 percent of annual precipitation in most areas and concentrated in the four months of June to September.[20] While two-thirds of China's land is threatened by floods and tropical typhoons ravaging coastal areas many times a year, hinterland droughts occur almost every year in the dry season. As China's history has recorded the world's top five deadliest floods and landslides of all times,[21] how to tame major rivers and lakes and handle famines related to floods and droughts has always been a life or death issue for each dynasty, which had to set aside the bulk of the country's material and human resources to building water conservation works and disaster relief systems.

As early as in 1950, one year after the PRC was founded, the government put in place a specific bureaucracy called the Central Flood Control Headquarters (*zhongyang fangxun zongzhihuibu*) to oversea and coordinate related departments in coping with severe floods, with Vice Premier Dong Biwu being the Director. In 1971, this organization was restructured into the Central Flood Control and Drought Relief Headquarters, which resumed its original form of the Central Flood Control Headquarters in 1985. In 1992, the State Council issued a circular to formally rename the organization as the State Flood Control and Drought Relief Headquarters (*guojia fangxun kanghan zongzhihuibu*),[22] which according to the PRC Flood Control Law has the responsibility of leading and organizing the anti-flood and drought relief work at the national level. The State Flood Control and Drought Relief Headquarters is usually headed by a vice-premier-level official with the Minister of Water Resources and Deputy Secretary-General of the State Council serving as two deputy directors of the Headquarters.[23]

Compared with the China National Committee for Disaster Reduction, which functions as an inter-agency mechanism dealing with all kinds of natural disaster relief work in general, the State Flood Control and Drought Relief Headquarters is a specifically designated mechanism that coordinates anti-flood and drought relief work among various ministries and departments. Due to such differences in functions and responsibilities, the Ministry of Civil Affairs plays a central role in the mechanism of NCDR, while the Ministry of Water Resources is the pivotal department in the mechanism of the State Flood Control and Drought Relief Headquarters. The two systems overlap sometimes, especially when the country is fighting against disastrous flooding like the two Yangtze River floods in 1998 and 2010. Under such circumstances, the State Flood Control and Drought Relief Headquarters usually becomes the dominant mechanism in which the Ministry of Water Resources instead of the Ministry of Civil Affairs functions as the pivotal sector among its peers.

The Chinese government set up another *ad hoc* inter-agency mechanism called the Earthquake Relief Headquarters of the State Council (*guowuyuan kangzhen jiuzai zongzhi huibu*) in response to a magnitude-8.0 earthquake in southwest Sichuan Province on May 12, 2008, the deadliest quake in China in the reform period. During Premier Wen Jiabao's flight to Sichuan just 90 minutes after the earthquake, he announced the establishment of the Earthquake Relief Headquarters, which was headed by him and comprised different ministries, departments, and military sectors. On May 18, 2008, the State Council publicized the component details of the Earthquake Relief Headquarters, which was divided into nine inter-agency workgroups to coordinate the related work among various sectors.[24] A ministry or department was appointed to play a pivotal role in each group, with the General Staff Department of the Chinese army being the coordinator in the emergency response group, the Ministry of Civil Affairs being the coordinator in the disaster relief group, the China Earthquake Administration being the coordinator in the earthquake monitoring group, and the Ministry of Health being the coordinator in the public health and anti-epidemic group.

The Earthquake Relief Headquarters of the State Council was specifically set up to deal with the 2008 Sichuan Earthquake, but such an *ad hoc* mechanism continues to exist long after the occurrence of the quake mainly because the reconstruction and recovery process needs to take a long time in the province, and the central government has realized the importance of maintaining the *ad hoc* mechanism to respond to severe earthquakes that may strike China again in the future. When Yushu, northwestern China's Qinghai Province, was hit by a devastating magnitude 6.9 earthquake in April 2010, the Earthquake Relief Headquarters of the State Council immediately organized emergency aid and disaster relief work at the national level.

Both the State Flood Control and Drought Relief Headquarters and the Earthquake Relief Headquarters are *ad hoc* inter-agency mechanisms under the State Council, but a structural distinction between them is that the Earthquake Relief Headquarters is a multi-layer organization comprising various workgroups, each of which is headed by certain ministry or department, while the State Flood Control and Drought Relief Headquarters has no such subgroups and is simply a double-layer organization in which the Ministry of Water Resources plays a pivotal role. Besides these two mechanisms, the lower-rank State Headquarters for Forest Fire Control was established in May 2006 to coordinate related work among various sectors, with the director of the State Forestry Administration being the commander-in-chief.[25]

The evolution of operational mode

In the past thirty years, the Chinese government has made significant changes in how to operate its natural disaster management system. During the reform period in which the PRC has been paying more attention to the institutionalization and legalization of government activities, the country has enacted more than thirty laws and regulations related to disaster management, including the PRC Flood Control Law, the Law on Earthquake Preparedness and Disaster Reduction, the Emergency Response Law, the Regulation on Forest Fire Control and the Regulation on the Participation of Armed Forces in Emergency and Disaster

Rescue Actions. The Chinese government has enacted detailed regulations on the comprehensive disaster relief work, the reporting and monitoring of disaster situation, the management and use of aid materials and fund as well as on how to receive international and domestic donations. Under the related laws and regulations, the Chinese government has institutionalized an emergency response system comprising the National Master Plan for Responding to Public Emergencies, five national thematic disaster response plans and emergency response plans for 15 central government departments, and their detailed implementation plans and operation norms.[26] Compared with the PRC's early experiences in disaster management, the bureaucratic system in the reform period, facilitated by all these laws, regulations, and plans, gradually had a clarified assignment of duties among major departments and standardized the procedures through which the government, the armed forces, businesses, and civil society plan for and reduce the impact of disasters, react during and immediately following a disaster, and take steps to recover after the disaster.

Having drawn lessons from past experiences, the Chinese government has paid close attention to the disaster preparedness and prevention phase, with detailed thematic disaster response plans enacted and enormous funding and resources put into precaution measures in the recent two decades. In the pre-reform period, China had no specific emergency response plan to deal with certain type of natural disasters, and neither did China classify the disasters according to their severity, controllability, and affected areas. Now, besides the National Emergency Response Plan for Natural Disaster Relief, which is applicable to relief actions for all kinds of natural disasters, *ad hoc* response plans including the National Emergency Response Plan for Flood and Drought Disasters, National Emergency Response Plan for Earthquake Disasters, National Emergency Response Plan for Geological Disasters, and National Emergency Response Plan for Very Severe and Most Severe Forest Fire Disasters are also in place to handle specific types of disasters. According to the severity and controllability, the Chinese government has classified natural disasters and other emergency events into four response categories under the National Master Plan for Responding to Public Emergencies: (a) Level I for the most severe situation, where the response is organized directly by the central government, together with the affected provincial and local governments; (b) Level II for very severe events, where the relevant provincial governments are primarily responsible for organizing responses, with the assistance of the central government; (c) Level III for severe events; and (d) Level IV for general public emergencies, where events could be dealt with mainly by local governments, with the assistance of relevant central government departments.[27] Through the classification of natural disasters based on severity, the central government has found it easier to figure out the amount of funding that should be allocated to different disasters, as well as which level of government should play a dominant role in disaster relief work.

Since local governments in China had no power to raise revenue and most of their expenditures were disbursed by the central government in a strict top-down system prior to 1980, the central government had to be responsible for allocating the relief fund to any place that had encountered natural disasters. In the 1980s and early 1990s, although local governments could gradually arrange their own revenue and expenditure under a contractual fiscal system that was characterized by defining a fixed sum paid to or received from the central government, most of the disaster relief fund still came from the central government. From 1983 to 1994, 85 percent of the disaster relief fund used in the Anhui Province was allocated by the central government, while only about 15 percent was from the provincial and county-level governments.[28] After China in 1994 introduced the Tax Sharing System with which local governments were allowed to share tax revenues with the central government, local governments at all levels also started to disburse more expenditure on disaster management. Local governments today shoulder about 30 percent of China's total fiscal expenditure on disaster management, which is much higher than the ratio in the pre-1994 period. Yet they are still being pressured to pay more for disaster relief so as to ease the fiscal burden upon the central government.

Shifted from its insular attitudes towards foreign aid for disaster relief in the Cold War, China has been readily receiving all kinds of international donations through different channels in the reform period. The Ministry of Civil Affairs, the Ministry of Foreign Affairs as well as the non-governmental Red Cross Society of China and China Charity Federation are the major sectors involved in receiving overseas aid and donations. After the occurrence of the 2008 Sichuan Earthquake, the Chinese government not only accepted international aid and donations, but also welcomed rescue and medical teams from foreign countries like Britain, France, Indonesia, Russia, and Singapore to the earthquake regions.

Norms and learning in China's natural disaster management

As a country vulnerable to many types of natural disasters, China in the past thirty years has made tremendous effort to improve its complicated bureaucratic structure in managing those disasters and to enhance relevant financial and technical preparation. Meanwhile, thanks to the opening-up process and increased international engagement, new norms and guidelines have been gradually introduced to China's natural disaster management and many of them have become well-established principles guiding the country's disaster relief work in the new century. As a result, the Chinese government has managed to control the casualties caused by various kinds of natural disasters at relative low levels[29] and its achievement in this regard has been acknowledged by the international community.[30]

Established norms

1. Prioritizing prevention and preparation measures

Just like many other parts of the world, China today has been making disaster mitigation and preparedness its priority task at policy, institutional, and operational levels. In the disaster management cycle, pre-disaster mitigation and preparedness activities not only reduce the probability of disaster occurrence or the negative effects of unavoidable disasters, but also achieve a satisfactory level of efficiency in responding to any emergency situation. Therefore, compared with the post-disaster phases of response and recovery, the preventive and preparation measures are more cost-effective and thus deserve intensive policy attention.

Among all the preventative measures, building up flood-control facilities and related warning/forecasting systems has drawn the most attention from the Chinese government, especially since the Yangtze River floods in 1998, the heaviest in the reform period, which led to a series of levee collapses and killed more than 4,150 people. In the early stage of the reform period when the state followed a blatantly urban-biased development strategy with most resources diverted to glamorous industrial and urban projects instead of water conservancy and forestation projects, China had to bear the inestimable environmental costs of soil erosion and deforestation that exacerbated the floods (Figure 7.1 shows that areas affected by floods had been on the rise from 1978 to 1998). Since 1998, however, substantial funds have been generated both from central and local governments for the reinforcement of the levees and for building the flood storage and detention basins. China has invested enormously in the world's largest flood control project, the Three Gorges dam that was completed in 2006. From 1998 to 2008, the total areas affected by floods in China were greatly reduced (Figure 7.1). The government has also set up nationwide early-warning networks to monitor and identify risks and threats of meteorological, oceanic, geological, earthquake, and forest fire disasters.

With respect to preparedness measures, the Chinese government has not only mapped out national emergency response plans for major types of natural disasters, but also built up a three-tiered manpower mobilization mechanism that includes professional teams, volunteers, and community-based human resources to respond to disasters. Besides, the Chinese government has established special financial budgets for the response to and relief of general disasters, and several central and local government departments including civil affairs, water resources, agriculture and forestry have established a system for stocking disaster relief materials, which comprises ten central government stocks and local stocks in areas where disasters happen frequently.[31]

2. Clear assignment of duties within the bureaucracy

Local governments have been authorized by the National Emergency Response Plan for Natural Disaster Relief to play a larger role in disaster relief work. The plan articulated the principle of "combination of central and regional system while giving priority to the regional system" (*tiaokuai jiehe yikuai weizhu*), requiring provincial, municipal, county-level, and township governments to take due responsibility of implementing local emergency response plans, moving disaster-hit people to safer places and reporting and

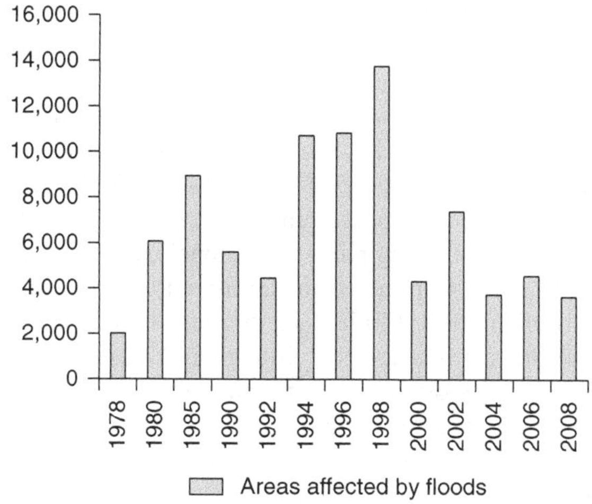

Figure 7.1 Areas affected by floods in China (1,000 hectares).

Source: National Bureau of Statistics, *Guojia tongji nianjian 2009* [*China Statistical Yearbook 2009*] (Beijing: Zhongguo tongji chubanshe, 2010), p. 479.

assessing disaster situation.[32] After the Tax Sharing Reform in 1994, which greatly enhanced the fiscal capacity of local governments at various levels, regional governments have been responsible for a larger share of disaster relief expenditures. Yet in contrast to the sharp increase of their revenues, local governments' payment growth for disaster management is still sluggish.

Horizontally, the central government has set up various inter-agency mechanisms to ensure the clear-cut division of work among involved ministries and departments. For routine jobs on disaster relief, it is the Ministry of Civil Affairs that plays the pivotal role in coordinating the work of different sectors, while in catastrophic disasters like the 2008 Sichuan Earthquake or the 2010 Yangtze River Floods, such *ad hoc* inter-agency panels as the Earthquake Relief Headquarters or the State Flood Control and Drought Relief Headquarters command the disaster relief campaigns. According to relevant legal documents, all the involved departments and sectors have been assigned well-defined jobs in the colossal bureaucratic system on disaster management to avoid administrative chaos, with punishment in place for the breach of duty.

3. Openness to international aid and media

One of the sharp contrasts between the disaster relief work in the Maoist era and the reform period is China's gradual openness to international aid and media coverage. Many blamed the Chinese government's rejection to foreign aid for the exacerbation of the destructive consequence of the Tangshan quake in 1976. Meanwhile, before the reform period, the PRC government had regarded the death tolls in natural disasters as state secret and refused to release them to the public. Foreign media were not allowed to cover the disaster relief work on the spot at that time.

In the early 1980s the reformist leadership decided to accept foreign aid and opened its disaster relief system to the outside world, with the United Nations as its major institutional channel to engage the world. The year 1989 was a turning point when China joined the International Decade for Natural Disaster Reduction (IDNDR) initiated by the UN and set up the National Commission for the IDNDR, which later became the top-level decision-making panel on national disaster relief work. As the predecessor of the National Committee for Disaster Reduction, the National Commission for the IDNDR in the 1990s had learnt a lot of modernized norms and practices in the field of disaster management through all kinds of international exchanges and activities. As a move to greater transparency, China in 2005 announced that it would no longer treat the death toll in natural disasters as a state secret, with the National Administration of State Secrets declaring the declassification of disaster-related death tolls almost at the same time.[33]

During the Sichuan Earthquake in 2008, the country's worst natural disaster in the reform period, official media provided round-the-clock coverage of the earthquake and the rescue efforts, demonstrating remarkable openness. The Chinese government not only accepted international aid and donations, but also welcomed rescue and medical teams from foreign countries to the earthquake regions.

4. Deploying armed forces in time

The PRC has frequently mobilized troops to combat severe disasters like floods, earthquakes, snowstorms, and forest fires. Now with 2.3 million soldiers, the People's Liberation Army (PLA) has been institutionalized as the country's most important rescue team in major disaster relief campaigns. According to the PRC Law on National Defense that was enacted in 1997, military personnel on active duty shall "accomplish the tasks of dealing with emergencies and providing disaster relief."[34] During a rare snowstorm that affected much of southern and central China in early 2008, 306,000 PLA soldiers plus 1.07 million militia and army reservists were mobilized in the disaster relief effort.[35] Within hours of the Sichuan Earthquake in 2008, thousands of soldiers and policemen were being deployed, with military helicopters heading for the epicenter to help with rescue efforts. Within three days, tens of thousands of troops forged into the quake regions to pull people from the rubble.

5. Promoting advanced technology and know-how

China today is much better equipped technologically than before to deal with natural catastrophes, with improved high-tech systems for forecasting and monitoring disasters. Since the effectiveness of disaster management relies greatly on the utilization of related technology and information, the country has made strenuous efforts to improve its knowledge and technology capacity in all fields related to disaster mitigation. Relying on meteorological, oceanic, and land resources satellites, as well as ordinary airplanes and unmanned planes, China has been using remote sensing and satellite positioning technology for the surveillance tasks. Within two hours of the Sichuan Earthquake, a map indicating basic information about the epicenter and major affected areas had been prepared by the National Disaster Reduction Centre, an information-sharing platform under the Ministry of Civil Affairs. In the following days, 120 maps and reports derived from satellite and airplane images were submitted by the Centre and its cooperative partners. Manned and unmanned airplanes equipped with remote sensors flew over the areas hit by the quake to collect field information with a view to more effectively deploying rescue and mitigation forces and relocating affected people.[36]

Problems and lessons

Despite the remarkable progress made by the PRC in natural disaster management, problems still exist and through learning lessons from previous experiences, the country can further improve its relief work to reduce potential losses in disasters.

Currently the government is still playing an overwhelming role in disaster management, with non-governmental organizations (NGOs) unable to provide sufficient assistance. Although indigenous NGOs, a symbol of the emerging civil society, have become increasingly visible players in China's disaster relief work, their roles and functions are still quite restrained under China's current state-centric political systems, and therefore, they are not as strong and influential as their international peers. The Sichuan Earthquake in 2008, however, witnessed an unprecedented civic mobilization facilitated by Internet communications. Immediately after the quake, NGOs and citizens around the nation responded to the disaster just as quickly as the government. The day after the earthquake, several environmental and educational NGOs in Beijing initiated a "Green Ribbon" campaign. Their members and volunteers fanned out in the streets in fund-raising and blood donation drives. On the same day, 57 NGOs issued a joint statement calling for concerted disaster relief efforts among all NGOs, and 51 other civic groups jointly established an office in Chengdu to coordinate NGO relief activities.[37]

Due to the government's strict control and restrictions over NGOs plus their weak capacity building, most NGOs except government-organized NGOs (GONGOs) like the Red Cross Society of China and China Charity Federation could only play a very limited role in the relief work. With the Sichuan Earthquake being an exception, most NGOs in China failed to take noticeable actions in response to such major disasters as the snowstorm in 2008 or the deadly floods and mudslides in 2010. Past experiences in western countries show that significant supervision and participation from NGOs and other civil society groups are requisites for better disaster management, so if China wants to address natural disasters more efficiently, it has to improve the institutional environment for the vibrant growth of NGOs.

The central government is now overloaded with excessive responsibilities in disaster management, having to take care of almost all trivialities related to organizing donations, sending reports to the disaster zones, transporting aid materials, and dispersing compensation funds. This increases the difficulty in coordinating work among various sectors and in implementing detailed instructions and policies, and thus on many occasions causes inefficiency and even corruption. The one-party political system, which lacks sufficient supervision from opposition parties and independent media, also exacerbates embezzlement of donation funds and materials. Local governments at various levels sometimes do not follow their superior authorities' instructions when allocating relief and recovery funds. As for post-disaster compensation, insurance companies still play a minor role due to their limited population coverage, while the Chinese government is responsible for most of the payment.

Obsessed with the high-speed economic growth that has become a major indicator to measure administrative chiefs' performance, local governments at various levels often pay much more attention to glamorous industrial and urban projects instead of water conservancy and forestation along small and medium-sized rivers. Due to the insufficient funding from the local budgets, some remote seismological observatories and disaster reduction facilities cannot maintain normal functions. The earthquake in 2008 and floods and mudslides in 2010 can be considered timely wake-up calls for the government on the need to pay more attention to the development of local disaster reduction infrastructure that has lagged behind the expansion of urban areas and urban population growth.

Notes

1 See, for instance, Ye Yaoxian, "Chinese Experience with Post-Natural-Disaster Reconstruction," a paper prepared for Conference on Improving Post-disaster Reconstruction in Developing Countries, McGill University, Montreal (May 23–25, 2002), at http://www.GRIF.UMontreal.ca/pages/irecconference.html; Andrea Janku, "Towards a History of Natural Disasters in China: The Case of Linfen County," *The Medieval History Journal,* Volume 10, Issues 1&2 (2007), pp. 267–301; and "Chapter 1: Natural Disasters in China," *China's Actions for Disaster Prevention and Reduction*, announced by Information Office of the State Council of the People's Republic of China on March 11, 2009.

2 Karl Wittfogel, an eminent German-American historian, dubbed China as a kind of "hydraulic civilization," which refers to the culture having an agricultural system that is dependent upon large-scale government-managed waterworks – productive (for irrigation) and protective (for flood control). In his book *Oriental Despotism* (New Haven, CT: Yale University Press, 1957), Wittfogel believed that such civilizations in the Orient were quite different from those of the West.

3 Early historians have dated China's management of constant flooding along the Yellow River back to the 20th century BC, when *Da Yu* was asked by Emperor Yao to construct dams and dykes along the Yellow River to control floods. See Linda Liang, "Water Management and Allocation of the Yellow River Basin," a report to the International Water Management Institute (IWMI), (2005), p. 1, available at http://www.iwmi.cgiar.org/assessment/files_new/research_projects/River_Basin_Development_and_Management/Liang_WaterPolicy_YRB.pdf.

4 Early literature on the *huangzheng* research can be dated back to the South Song Dynasty (AD 1127–1279), when Dong Wei wrote his influential book *Book on Disaster Relief and Lifesaving* (*Jiuhuang huomin shu*). Other representative works on the same topic in the modern times include Deng Tuo, *Zhongguo jiuhuang shi* (China's History of Disaster Relief) (Shanghai: The Commercial Press, 1937); and Meng Shaohua, *Zhongguo zaihuangshiji* (China's History of Disasters and Famines) (Beijing: Zhongguo Shehui Chubanshe, 1999).

5 Sun Shaocheng, *Zhongguo jiuzai zhidu yanjiu* (Research on China's Disaster Relief Institutions) (Beijing: The Commercial Press, 2004), p. 54.

6 Dong Wei (South Song Dynasty), *Jiuhuang huomin shu* [*Relief and Lifesaving, The Third Volume of Book on Disaster*], at http://zh.wikisource.org/zh/%E6%95%91%E8%8D%92%E6%B4%BB%E6%B0%91%E6%9B%B8.

7 Dawen Yang and Hiroshi Ishidaira, "Profile of the Yellow River Basin," in Tetsuya Kusuda (ed.), *The Yellow River: water and life* (Singapore: World Scientific, 2010), p. 17.

8 Sun, *Zhongguo jiuzai zhidu yanjiu*, pp. 83–84.

9 Ibid., p. 55.

10 Gong Shuduo, *Zhongguo shehui tongshi-mingguo juan* (Chinese Social History-ROC Volume) (Taiyuan: Shanxi Education Press, 1996), pp. 500–504.

11 Yang and Ishidaira, "Profile of the Yellow River Basin," p. 18.

12 Sun, *Zhongguo jiuzai zhidu yanjiu*, p. 130.

13 Kallie Szczepanski, "The Great Tangshan Earthquake of 1976: The Natural Disaster That Ended China's Cultural Revolution," at http://asianhistory.about.com/od/china/a/TangshanQuake.htm (accessed 1 November 2010).

14 Sun, *Zhongguo jiuzai zhidu yanjiu*, p. 138.

15 Chen Gang, "China's Floods in 2010: A Stress Test for State Capacity," *EAI Background Brief*, No. 553 (August 2010), pp. 5–6.

16 "Guowuyuan bangongting guanyu yinfa minzhengbu zhineng peizhi neishe jigou he renyuan bianzhi guiding de tongzhi [Circular on the Departments Established Inside the Ministry of Civil Affairs and Related Established Posts]," announced by State Council General Office, No. 60 (1998) on June 17, 1998, at http://law.lawtime.cn/d466727471821.html.

17 Asian Disaster Reduction Center (ADRC), "Information on Disaster Risk Reduction of the Member Countries: China," at http://www.adrc.asia/nationinformation.php?NationCode=156&Lang=en&NationNum=22 (accessed on November 1, 2010).

18 "Function of the Disaster-Relief Bureau of the PRC Ministry of Civil Affairs," on the Official Website of the Ministry of Civil Affairs at http://jzs.mca.gov.cn/article/jgzn/ (accessed on November 1, 2010).

19 "Guojia jianzai zhongxin (National Disaster Reduction Center)," on the Official Website of the Ministry of Civil Affairs, at http://www.mca.gov.cn/article//zwgk/jggl/jgzn/200712/20071200006019.shtml (accessed November 2, 2010).

20 Sometimes more than 70 percent of regional rainfall is concentrated in the two months of July and August. See Chen Gang, "China's Floods in 2010: A Stress Test for State Capacity," p. 5.

21 Estimated deaths in the 1931 China floods, the world's deadliest floods and also the deadliest natural disaster of all time, were between 2 million and 4 million. The 1887 Yellow River flood ranked second in death toll in both floods and natural disasters, claiming lives of between 0.9 million to 2 million. The 1938 Yellow River flood ranked third, with 500,000–700,000 deaths.

22 "Brief Introduction of the State Flood Control and Drought Relief Headquarters," on the Official Website of the Office of the State Flood Control and Drought Relief Headquarters, at http://sfdh.chinawater.com.cn/jgsz/jgsz.htm#2 (accessed on November 2, 2010).

23 Since 2003, Vice Premier Hui Liangyu has been Director of the State Flood Control and Drought Relief Headquarters, whose member departments include the Ministry of Water Resources, Ministry of Finance, Ministry of Public Security, Ministry of Civil Affairs, Ministry of Land and Resources, Ministry of Transport, Ministry of Agriculture, Ministry of Health, China Meteorological Administration, Ministry of Railway, Ministry of Commerce, State Administration of Radio Film and Television, National Development and Reform Commission, Ministry of Industry and Information Technology, Ministry of Housing and Urban-Rural Development and State Administration of Work Safety. For relevant information, please refer to http://fxkh.mwr.gov.cn.

24 "Guanyu guowuyuan kangzhen jiuzai zongzhi huibu gongzuozu zucheng de tongzhi [Circular on the Component Workgroups of the State Council Earthquake Relief Headquarters]," *Xinhua News*, May 18, 2008.

25 "guanyu chengli guojia senlin fanghuo zhihuibu de tongzhi [Circular on the Establishment of the State Headquarters for Forest Fire Control]," announced by State Council General Office, No. 41 (2006) on May 29, 2006, at http://www.gov.cn/zwgk/2006-05/31/content_296311.htm.

26 The United Nations Economic and Social Council, "Implementation of the Hyogo Framework for Action in Asia and the Pacific: Case Study: The National Disaster Management System of China and Its Response to the Wenchuan Earthquake" (December 2008), note by the Secretariat presented at the Committee on Disaster Risk Reduction, Economic and Social Commission for Asia and the Pacific, Bangkok (March 25-27, 2009), p. 9, available at http://www.unescap.org/idd/events/cdrr-2009/CDR_2E.pdf.

27 The United Nations Economic and Social Council, "Implementation of the Hyogo Framework for Action in Asia and the Pacific," p. 9.

28 Sun, *Zhongguo jiuzai zhidu yanjiu*, p. 192.

29 From 1999 to 2005, 2,555 people were killed in natural disasters annually in China. See "Minzhengbu gongbu 2006nian quanguo yin ziran zaihai siwang renkou qingkuang [The Ministry of Civil Affairs Publicizing the Death Toll Caused by Natural Disasters in 2006]", on the official website of the Chinese government (January 4, 2007), http://www.gov.cn/zhibo46/content_490836.htm.

30 See, for example, "U.N. Leader Praises China's Quake Response," *The New York Times*, May 25, 2008, http://www.nytimes.com/2008/05/25/world/asia/25china.html; The United Nations Economic and Social Council, "Implementation of the Hyogo Framework for Action in Asia and the Pacific."

31 The United Nations Economic and Social Council, "Implementation of the Hyogo Framework for Action in Asia and the Pacific," p. 14.

32 For details of the principle of "combination of central and regional system while giving priority to the regional system," see Article 6 (Emergency Response) of the *Guojia ziran zaihai jiuzhu yingji yu'an* [*National Emergency Response Plan for Natural Disaster Relief*] (January 2006), at http://www.gov.cn/yjgl/2006-01/11/content_153952.htm.

33 "China's State Secrets Agency Will Guard One Less: Death Tolls," *The New York Times*, September 13, 2005, at http://www.nytimes.com/2005/09/13/international/asia/13china.html.

34 Article 58 of *the PRC Law on National Defense* (1997).

35 "306,000 troops mobilized to combat snow disasters," *Xinhua News*, February 3, 2008.

36 The United Nations Economic and Social Council, "Implementation of the Hyogo Framework for Action in Asia and the Pacific," p. 6.

37 Yang Guobin, "Sichuan Earthquakes and Relief Efforts: The Power of the Internet," *EAI Background Brief*, No. 389 (June 25, 2008), pp. 5–6.

Index